A short history of pottery raw materials
and the Cheddleton Flint Mill

Robert Copeland

A short history of pottery raw materials and the Cheddleton Flint Mill

Introduction

A short history was written in 1968 for the North Staffordshire Journal of Field Studies and published in Volume 9, 1969. In 1971, the Cheddleton Flint Mill Preservation Trust reprinted it with extra illustrations as a separate booklet; this has proved to be so popular that the author has revised and enlarged it, drawing on recently published writings which are included in the expanded list of references.

The Trustees hope that this new edition will prove to be of help to all those interested in the study of industrial processes and in the historical development of raw materials for the pottery industry.

Acknowledgements

It has been usual for writers on pottery history to condemn Simeon Shaw for inaccuracies in his book. In this short history references to Shaw's History of the Staffordshire Potteries are augmented with references from earlier works. My grateful thanks are given to all previous authors, and especially to Norman Emery and his staff at the Hanley Reference Library. Also to the many friends who have helped to ensure that this work is as accurate and up-to-date as possible. Among these I include: Dr Trevor Ford, Dr W. L. German, Reginald Haggar, Nellie Kirkham, Jim Kelly, Alex Malkin, Arnold Mountford, Ted Royle, Frank Underwood, the British Ceramic Research Association Library staff and that of the British Museum.

Classification (Dewey)
738 (Pottery) ·12
622 (Milling) ·73 (Crushing and Grinding)

ISBN 950 2668 0 9

Published by the
Cheddleton Flint Mill Industrial Heritage Trust
Near Leek, Staffordshire
Designed by John Bedford MSIA
Printed in England by
Wood, Mitchell & Co. Ltd., Hanley, Stoke-on-Trent

A short history of pottery raw materials
and the Cheddleton Flint Mill

The origins of the Pottery Industry of North Staffordshire

Butter pots of red clay – made in Burslem c1650

The question is often asked: why is the pottery industry concentrated in North Staffordshire when so much of the raw materials come from the South West of England? In North Staffordshire there are unusually good deposits of clays, ranging from red firing clays to stoneware and refractory clays for withstanding high temperatures. Although good clays can be found in many parts of the country, the contributory cause in confirming the establishment of the industry in North Staffordshire was the availability of an excellent bituminous coal and its unusual geological formation of outcropping:[11] salt and lead – important glazing materials – were found in neighbouring counties.

By the time that Devon ball clays were being brought into the district, there was more than just a rural craft in Burslem: in fact, there was a substantial industry catering for market needs many miles distant. Moreover, a network of communications was well established, enabling materials to be brought in and finished goods carried out. The transport of the heavier materials used in bulk such as common and refractory clays and the all-important coal was localised; clearly, it was much cheaper to bring the body materials to the coalfield than vice versa, while the community would not be expected to leave their established homeland en masse for a distant part of the country. The importance of coal can be more easily understood when it is realised that from 17 to 20 tons of coal are needed to fire one ton of domestic clay-ware in a traditional bottle oven.

This book traces the background history of the pottery industry as it relates to raw materials and the early importance of power and better communications in North Staffordshire.

Local Clays and Sand

The development of the pottery milling industry has been closely linked with the series of improvements made in the composition of ceramic tableware clays or bodies. Before 1700, pottery was made from various clays dug in the northern districts of the Potteries coalfield in association with outcrop coal.[103] These materials were tainted with iron and other impurities and many potters strove to find clays which would burn whiter than those of their competitors.[125] At first, they achieved whiter bodies by introducing white sand or pounded gritstone from Mole Cop;[99] for years this source was supposed to be that place on the border with Cheshire from which millstones were quarried (now called Mow Cop). Although on Hargreaves' map[48] (1832) a place called Mole Cop is shown South of Great Fenton, it seems certain that the Mole Cop on the boundary with Cheshire was the original and principal source of sand for the potters:[86] a fine white sand of good quality can be obtained there still. Pitt states that pounded gritstone from Baddeley Edge and Mole Cop was mixed with the whitest clay from Devonshire:[100] clays from Devonshire were used contemporarily with the pipe-clays found locally which were in use earlier. Dr. Plot notes that Mole Cop is a source of millstones of a special sort favoured by corn millers,[106] but does not mention the use of sand nor gritstone in pottery. The improvement in pottery bodies was advanced greatly by the use of whiter-burning clays:[59] in the Potteries area they were found especially in Shelton, a village between Handley Green (Hanley) and Stoke.[103]

Clay tobacco pipe

But pipe clays, as these ivory-burning clays were called because of their use for making clay tobacco pipes for smokers, occurred in thin seams, and could not be used extensively. These clays were used as applied ornament on *drab ware*, or as a slip to lighten a dark coloured basic clay.[85] Similar pipe clays were brought from Derbyshire from the area around Brassington.[59] These early bodies based upon native clays and sand were called *crouch ware*.[118] The plastic clays from Devon and Dorset began to be imported into North Staffordshire in about 1720.[125] Also called *pipe clays* they soon earned the name of ball clays, because of the large lumps, or balls, which were dug out; these weighed about 35 lbs and three on each side of a pack-horse's pannier saddle was an average load.[110] (In Dorset, there is a belief that the term 'Ball Clay' derives from an abbreviation of the word 'Tubal', a special adze-like spade used for digging the clay).[31]

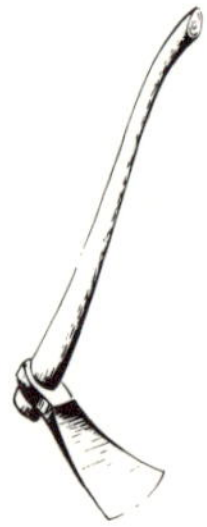

Tubal

The North Devon clay was shipped from the port of Bideford, the South Devon clay from Teignmouth[23] and the Dorset clay from Wareham.[54] The sailing ships unloaded at the port of Chester whence the clay was carried by pack animal or wagon to the Potteries; so, in the early days of the trade the clays were sometimes called Chester Clays. Later, some of the journey was made in barges along the River Weaver which was made navigable in 1733 as far as Winsford.[151]

The method of preparing clays for use was primitive. The Elers Brothers, who established a pottery just North of Newcastle-under-Lyme at Bradwell in about 1693 are said to have introduced methods of purifying clays for use by levigating them in water. This process would have allowed the coarser particles to sink to the bottom of the vessel, while the finer ones could be drawn off with the water into a sunpan where the water would be evaporated.[153] Thus the Elers Brothers could produce the fine textured red-ware suitable for turning on a lathe, and also for the relief decoration obtained either by impressing metal dies onto the surface of the turned ware,[87] or

Part of Hargreave's map showing Mole Cop south of Great Fenton. There are no indications of a road to it, nor any quarry. See also the *railway*, the tramway linking Lane End to the canal. Without this link it is doubtful if the south eastern towns would have developed in the way they did.

applying the clay sprig obtained from a mould, perhaps of pitcher (fired clay) or alabaster.[91]

During the excavations at Cheddleton Flint Mill for the foundations of a steam engine in March 1970, brick walls were discovered. An outer, circular wall of 9″ brickwork measured 16 feet diameter, while inside this were cross-walls forming what seems to have been nine separate arks or sections. Evidence of a white substance proved to be ground flint; it is suggested that, because the pan faced South, it was a sunpan – the first to be found in North Staffordshire.

This method of evaporating water was dependent on good weather for its success. In 1732, Ralph Shaw of Burslem is said to have invented the more effective process of drying his clay on a shallow kiln heated by flues running underneath.[126] This was perhaps the first advance on the primitive sunpan and made it easier for potters to work despite the lack of sunshine.[122] This method of driving off the surplus water is still used today in some of the most modern plants, adhering to the traditional practices because this method of drying yields improved characteristics which are important for certain raw materials.

Flint

It seems that John Astbury in or about 1720 was using Devon ball clay, first to improve the colour and texture of the body by applying a slip-dip or engobe, and shortly afterwards as an essential part of the pottery body itself.[100] It is

From Congleton
Rushton Spencer
Wolf Low
Earls Way house
Riding Side
Rushton
Overton
Stone Edge
Lee Side
Wolf Dale
Fairbrough
Biddulph Hill
Moor House
Whitmoor Wood
Bacon House
Biddulph
Gillow Heath
Hay Hill
Bradly Green
Braddocks Hay
Mole Cop
Red Cross
New Pool
Knypersley Hall
Park
Horton
Hunt House
Rudyard Hall
Rudyard
Black Wood House
Crowbrough
Harpers Gate
Harrades
Rowley Gate
Lime House
Lawton
Red Bull
Colmoor
Bullock House
Thursfield als New Chapel
Lane End
Brook Houses
Oldcote
Blackwood
Ladymoor Gate
Lane Head
Hole Houses
Hollins H.
Woodcock Hurst
Dunwood
Eudon
Bank End
Ladder Edge Common
Talk on the Hill
Talk Pitts
Golden Hill
Turn Hurst
Wedgwood
Great Chell
Little Chell
Ridgway
Wood House
Ball Green
Norton Green
Norton
Blakley Hall
Stanley
Stanley Head
Cheddleton
Shaffurlong
Dale House
Heath House
Brook House
Tunstal
Chatterley
Red Street
Bignole Hill
Bradwall
Chesterton
Brown hill
Bradly
Pinax
Fields
Bank House
Ford Green
Coal Pits
Down-field Side
Milton
Bagnal
Greenwood Hall
Rownall
Rownal
Wood End
New Mill
Hot Lane
Burslem
Rushton Grange
Birches Head
Abbey Hilton
Birches
Wood Head
Wetle Moor
Pare Tree House
Holditch
Whit Field
Wolstanton
Etruria
Cow Bridge
Handley Green
Bucknall
Eaves Lane
Abbey Mill
Blakeley
Wash Well
Werrington
Shepherds House
Cellar Head
Over Moor
New House
Knutton
NEWCASTLE
UNDER LYNE
Fowl Hay
Shelton
Trent Hay
Stoke Lane
Stoke
Botslow
Brook Houses
Ubbeley
Holly Greave
Bentley
Bury Hill
Fenton Low
Little Fenton
Moss Field
Upper Holme
Ash
Haywood Grange
Old Lodge
Pavillion
Grindley Hill
Lower Lane
Fenton Park
Penkhull
Boolton
Lane Delph
Adderly Green
Bolton Gate
Caverswall Common
Hardwick
Hall
Pen Fields
Bentaley
Lime
E
H
I
L
L
Great Fenton
Sideway
Park
Holly Bush
Lane End
Wood Houses
Park Hall
Weston Coyney
Cookshill
Caverswall
Clayton
Millstone Green
Butterton
North Wood
Handford Bridge
Handford
Longton
Blurton
Meer
Moor Gen
Hanchurch
Bank House
New Inn
Hem Heath
Normacot Grange
Grindly House
Acton
Trentham
Newstead
Cold Rulings
Bliht Bridge

Left: A detail from Yates map of Staffordshire (1775). Scale 1 in: $1\frac{2}{3}$ mile Water mills are shown as the one at Cheddleton is clearly marked

also stated[133] that in 1720 Thomas Astbury of Shelton, recognising the merits of ground flint, introduced it into the pottery body,[148] but the credit of being the first to introduce flint more properly might be given to John Dwight of Fulham.[88] He recorded in his notebook in 1698 "when giving the ingredients for a white body, 'calcined beaten and sifted flints will doe instead of white sand and rather whiter but ye charge and trouble is more.'"

An analysis shows that the white clay dip overlying an off-white body of a scratch-brown tankard dated 1723 does contain flint while the body does not. Flints are very heavy and costly to transport and prepare, so were not economical to introduce into the body itself until improved transport and mechanical assistance in grinding became available. Contemporary documents prove that dipped wares were always less expensive than those made wholly of pipe clays and flint. Besides this, it has been shown by a study of the inventories of potters of the time, that flints were being used earlier than in 1720. A partnership between John Fenton and Thomas Hill in 1719 was formed to make 'flintware'.[155]

Daniel Bird is believed to have been the first potter to ascertain the exact quantity of flint necessary to form a satisfactory earthenware body;[122] the amount of flint in an earthenware of today is about 32%. Flint whitened the body and reduced the risk of warping, but care in firing was needed to prevent the ware cracking.

Lead glazing

There were also many improvements in the method of glazing. Before the seventeenth century, lead ore had been used for glazing ware by placing the powdered ore on the surface of the piece;[104] Dr Plot records that this lead ore came from Lawton Park in Cheshire.[105] This is puzzling because there is no evidence to show that there are any deposits of lead in this area. The nearest source of lead would be Derbyshire: lead ore, galena, was mined over a wide area of the county adjoining the border with Staffordshire. The galena was crushed to a powder which became known as 'Smithum'.[58]

Salt glazing

The method of glazing with salt was patented by John Dwight in 1671 (Patent No. 164) in conjunction with his manufacture of stoneware.[81] It is clear that this patent, and its extension of 1684 was infringed by other potters even as early as 1673 and the salt glazing of stoneware must have been practised by Staffordshire potters by 1693.[82] The method involved firing the clayware in an oven to bright red heat (1200°C) when salt was shovelled into the fire mouths.[116] At this temperature the salt volatalized to form a vapour which settled on the red-hot clayware and, combining with the silica in the body, formed a semi-matt or shiny glaze. The salt would have been brought from Cheshire where rock-salt was discovered in 1670 at Marbury, near Northwich.[134] By 1762, about 1400 tons of salt (£5,000 tax payable) was being consumed annually in the glazing of one sort of stoneware.[75] (The Salt Tax was imposed in 1694 and continued in force at varying levels of taxation until the tax was withdrawn in 1825).[83]

Cream colour

The new imported materials of flint and ball clays allowed the increased manufacture of a completely cream coloured body. By 1740, this type of ware was 'much in request': it formed the basis for much salt-glazed ware, and later, lead glazed ware.[88] Enoch Booth is said to have introduced an improved method of applying a clear, transparent glaze in 1751.[160] This process involved mixing a lead compound, usually red or white lead, with water, and dipping the article into this fluid glaze.

It is impossible to assign clearly defined periods during which a process was practised. While an inventor of a new method or discoverer of a raw material might adopt the practice at once, other potters might retain the older ones for many decades either from choice or necessity. Indeed in some potteries the new and old methods might be practised side by side. So it was that, after the reintroduction of lead as a fluid glaze, salt glazing continued to be practised for 50–60 years afterwards.[89]

Cream-coloured earthenware, then, was being developed as the popular domestic tableware by the beginning of the 1750's, and it is from this period that the new impetus in the growth of the Staffordshire Potteries really began.[152] The demand for cream coloured ware with a lead glaze grew steadily and by the time Josiah Wedgwood had built the Etruria works the trade was well established. Wedgwood's contribution in this field of activity was to discover a reliable formula for the body and glaze after conducting a series of experiments on a scientific basis, and suffering many disappointments and heavy losses before the control of the firing was mastered.[72]

His reward came when Queen Charlotte in 1765 bought a service of this refined cream coloured earthenware and allowed it to be called 'Queens Ware'.[63]

Benson's patent

After the introduction of flint into the body, the very serious effects of silicosis upon the operatives quickly began to be felt; the method of grinding in those days would have been dry, probably in ordinary grist mills and later by stamp mills, driven either by wind or by water wheel.[93]

Queen's ware made by Wedgwood at the Bell Works, Burslem

Thomas Benson, of Newcastle-under-Lyme took out a patent in November 1726 for a method of preparing flint for making white pots.[69] This patent is often quoted incorrectly, so a facsimile of the patent, No. 487, republished in 1857 is reproduced on the facing page. This first patent employed a type of edge runner mill followed by grinding with large iron balls, and this was clearly unacceptable due to the introduction of iron into the body. Benson then took out a second patent, No. 536, on the 14th January 1732, which was a variation of the earlier one in that initially the flints were crushed by two large stones acting as edge runners and subsequently ground finely on stone pans using large stone balls. The two patents, therefore, are virtually alike except in the substitution of stone for iron, and both stressing the necessity of crushing and grinding wet in order to minimise the inhalation of silica dust. The 1732 patent was granted to Benson for the term of fourteen years: James Brindley who was a well-known millwright in Staffordshire, improved

A.D. 1726 . . . . Nº 487.

Preparing Flints for Making White Pots.

BENSON'S PATENT.

GEORGE, by the grace of God, &c., to all to whom these Presents shall come greeting.

WHEREAS our trusty and welbeloved THOMAS BENSON, of Newcastle-vnder-Line, in our County of Stafford, Engineer, hath by his peticon humbly represented vnto vs, that in Staffordshire and some other parts of our kingdom there is a manufacture carried on of making white pots, the chief ingredient of which is flint stone, and the method hitherto vsed vsed in preparing whereof has been by pounding or breaking it dry, and afterwardꝭ sifting it through fine lawns, which has proved very destructive to mankind, insomuch that any person ever so healthfull or strong working in that business cannot probably survive above two yeares, occasioned by the dust sucked into his body by the air he breaths, which being of a ponderous nature, fixes there so closely that nothing can remove it, insomuch that it is now very difficult to find persons which will engage in the business, to the great obstruccon and detriment of the said trade, which would otherwise, by reason of the vsefulness thereof, be of great benefit and advantage to our kingdom; that the peticoner has with great pains and expence invented and brought to perfeccon "AN ENGINE OR NEW METHOD FOR THE MORE EXPEDITION WORKING THE SAID FLINT STONE, WHEREBY ALL THE SAID HAZARDꝭ AND INCONVENIENCES ATTENDING THE SAME WILL EFFECTUALLY BE PREVENTED;" that he has at his own charge made severall experiments of his said Invencon, whereby he fully knows the same will answer the endꝭ above proposed in every particular to the manifest improvement and advantage of the said manufacture, and preserving the lives of many of our subiects imployed therein, and proposes to performe the same in manner herein-after described, viz_t:—The flint stones are first sprinkled with water, insomuch that no dust can arise to the hurt or damage of the workmen, then crushed as small as sand by two large wheels of the bigness and shape of mill stones, made of iron to turn round vpon the edges by the power of a water wheel, and afterwardꝭ conveyed into large pans made of iron for that purpose circular, in which there are large iron balls, which by the power of the water wheel above named are driven round by such a swiftness of motion that in a little time the flint stones so broken are made as fine as oyl itself, without the vse of lawns, and when so done by turning of a cock empties itselfe into casks provided for that purpose, and so kept therein for the vses above menconed; and having humbly prayed vs to grant him our Royal Letters Patent for the said Invencon according to the Statute in such case made and provided, wee, being willing to give encouragement to all arts and invencons which may be of publick vse and benefit, are gratiously pleased to gratify him in his request.

Salt glazed whiteware plate containing flint

Map indicating the principal sources of raw materials

on Benson's methods by enlarging the diameter of the pan and by pushing large silica stones over the silica pavement, instead of using the local stones.[70] This improved design is the basis on which the pan grinding of pottery materials has rested until recently.

China clay and Cornish stone

The next major step was the discovery that Cornish stone and china clay would form a porcelain body. William Cookworthy, a native of Kingsbridge in Devon, after training in London set-up as a chemist and druggist in Plymouth,[30] and discovered these materials in Cornwall in about 1747 subsequently publishing his patent Porcelain formula in 1768:[9] however these two important materials were well known and were obtainable in modest quantities as early as 1759,[72] so the extension of Cookworthy's Patent in 1775[29] by Richard Champion with its clause permitting the use of china clay and Cornish stone for uses which did not infringe the formula of Cookworthy was probably only acknowledging legally a state of availability that had existed for some time.

The unrestricted use of these materials was not allowed until 1796. The addition of these two materials, both derived from granite in various stages of decomposition, was a tremendous step forward in the development of strong white tableware bodies but, of course, it required the stone to be ground in a similar way to flint.

Transport routes

These materials, together with flint and ball clays, were heavy and required bulk transport facilities to convey them from the ports to the Potteries district. The loads were transported along river navigation as far as was possible, the main routes to the North Staffordshire Potteries being:[157]

1 the River Severn up to Bridgnorth or Bewdley, from Bristol
2 the River Trent to Willington from Hull
3 the River Weaver to Winsford
4 the River Dee to Chester

From the most upstream wharf the materials were carried overland on pack animals: carts and waggons were used very little due to the bad condition of the roads.[76] Sometimes broad-wheeled waggons were used which permitted the wheels to ride over the heavily rutted surface of the appalling roads.

Packhorse and drover – the main form of transport

The Trent and Mersey canal

It is significant that the first major canal to be planned was through Staffordshire and was completed in 1777, after James Brindley had cut the Trent and Mersey Canal and had later surveyed a branch canal – the Caldon – from the Potteries down to Froghall. The significance lies in the marketing and commercial astuteness of the leading potters, especially Josiah Wedgwood and his friend Thomas Bentley, who urged forward the Trent and Mersey canal plan so that their trade would expand to meet the growing need for pottery. This plan for the Grand Trunk Canal was the backbone of James Brindley's

Manchester
Liverpool
River Mersey
Runcorn
Route 3
Northwich
Chester
River Dee
Winsford
Route 4
River Weaver
Trent and Mersey Canal
Leek
Burslem
Cheddleton
Stoke
Newcastle
Lane End
Stone
Sandon
Stafford
Bridgnorth
Route 1
Bewdley
Stourport

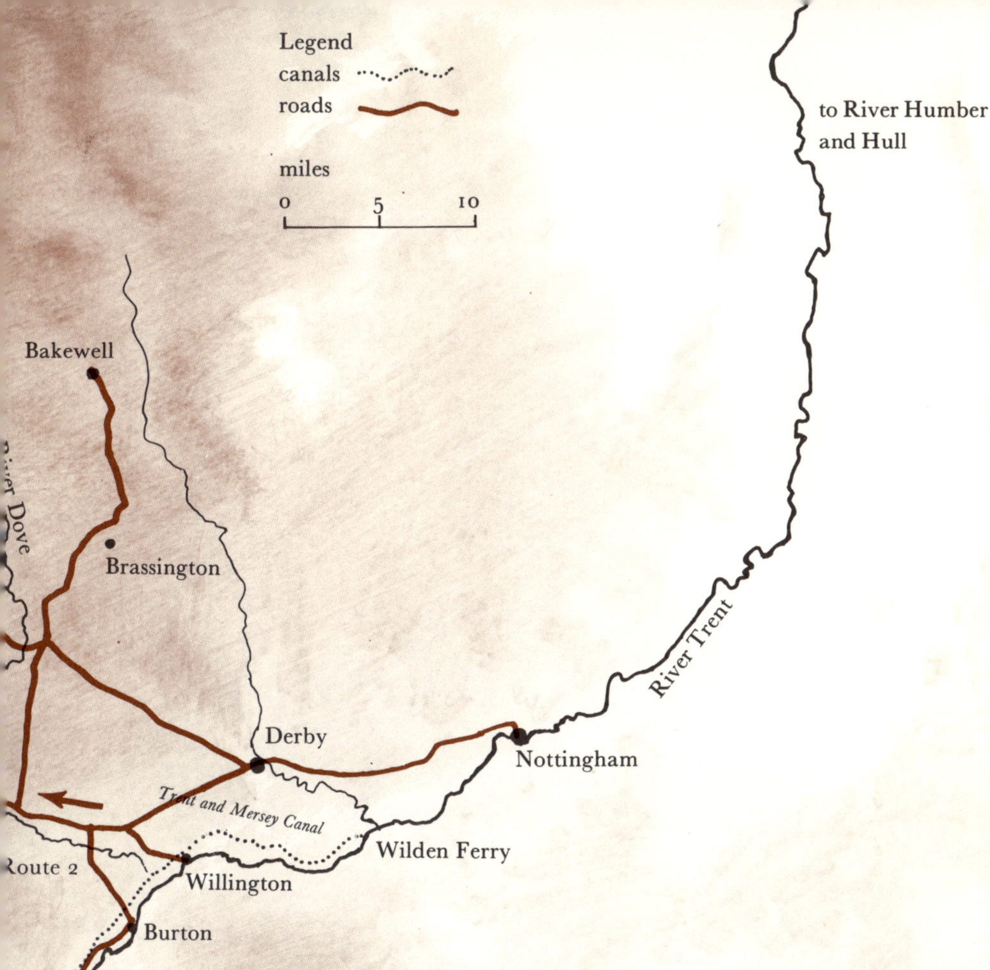

Wedgwood's factory at Etruria: the *Black Works*

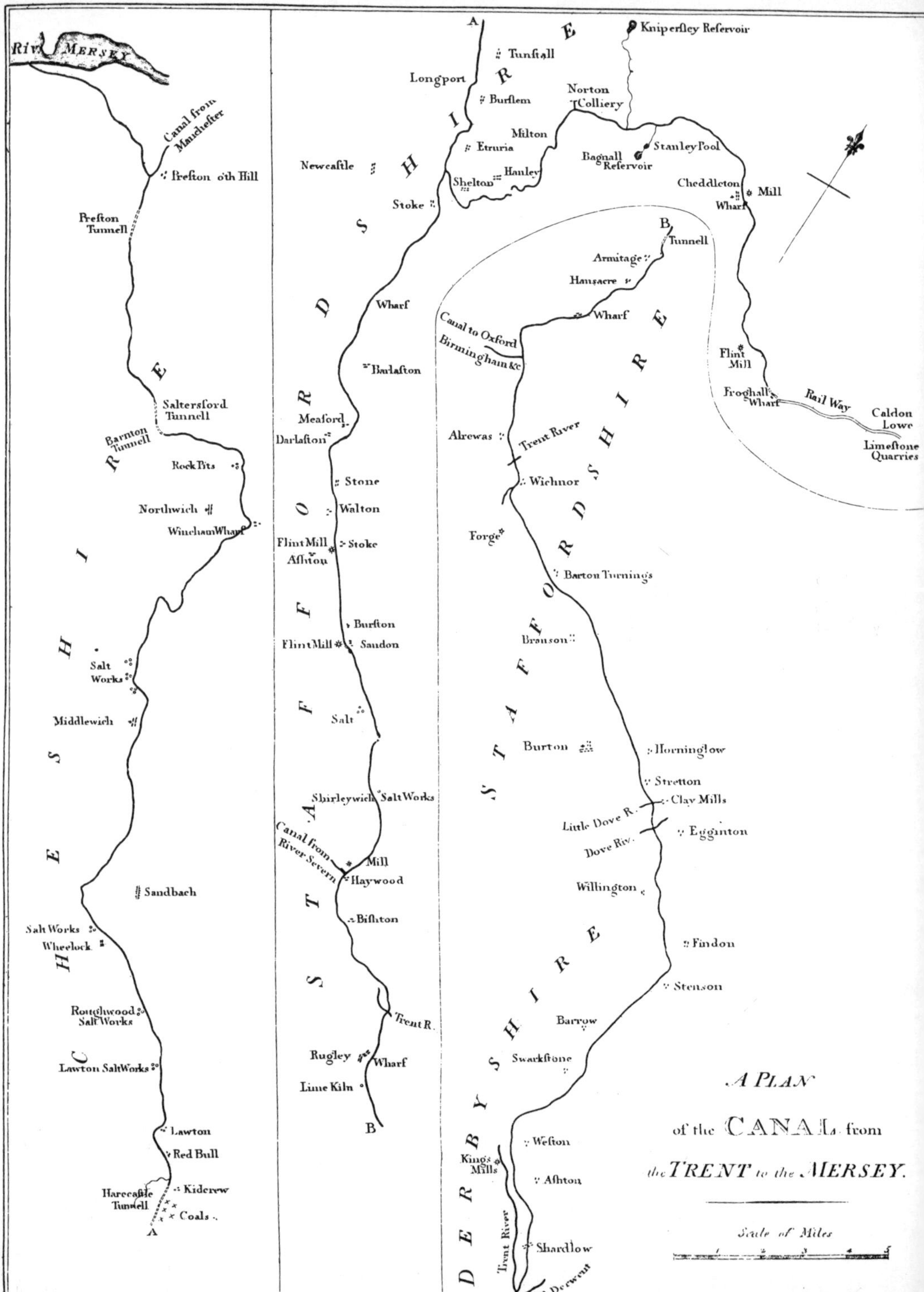

Published Feb.y 19. 1795, by I. Stockdale, Piccadilly.

Left: This plan by Aikin shows the importance of the canal in linking the flint mills with the Potteries[3]
Refer to the map on pp.10–11 to obtain a clearer idea of the orientation of the route

'Grand Cross' scheme for linking the major areas of the Midlands with a system of canals. It met with considerable opposition from the promoters of rival schemes,[78] but eventually the Bill was passed through Parliament in the Spring of 1766 receiving the Royal Assent on May 14th. The promoters worked fast: the first sod was cut at Brownhills, near Burslem on July 26th in the same year.[137] By 1772, the southern stretch of the canal from Stone was already opened,[90] and the whole length completed in 1777. The delay had been caused by the excessively difficult tunnelling at Harecastle. The Caldon branch (the Act for which was passed in 1776) was also opened in 1777. These two canals brought the Potteries nearer to two of the principal areas in which grinding by waterpower was taking place at that time, namely the Moddershall Valley near Stone, and the Churnet Valley. Amongst the Flint Mills in the former area was one built by Thomas Benson and John Gallimore with Astbury's son providing the capital.[113]

Expansion of trade

The pottery industry expanded very rapidly in the second half of the eighteenth century. The foundations of this expansion had been laid during the years 1720–1740, when flint and ball clay had been introduced into North Staffordshire to provide a white or cream coloured earthenware: the trade was helped further by the canalisation of the River Weaver in 1733,[158] and by the turnpiking of the road from Newcastle in North Staffordshire through Stoke and Lane End to Derby making easier the land journey to Willington, the head of navigation on the Trent.[140]

Some idea of the rate of growth can be gained from the level of shipments of ball clay (pipe clay) imported into the Potteries. Pipe clay taken upstream on the River Weaver increased from 150 tons in 1734, to 1300 tons in 1750 and to 1850 tons in 1760.[158] Exports of clay from Devon and Dorset and china clay from Cornwall are shown in Tables 1 and 2, overleaf. It will be noticed that the expansion of the trade leaps forward in the 1770's following the opening of the Trent and Mersey Canal, and this coincided with the American War of Independence and the Napoleonic Wars. These overseas military involvements reduced the imports of porcelain and faience from the Far East and Europe, not only because of the shortage of shipping available for such luxury goods but because of the import duty on porcelain[47] and of the increased patriotic support for British manufacturers' products, as well as a change in taste away from oriental designs.

Bone china

Lastly, but perhaps most important, came the successful introduction of bone into a porcelain body on a manufacturing scale by Josiah Spode before 1800.[161] Experiments in the use of bone in pottery making had been made for many years; Thomas Frye at Bow patented the manufacture of china using calcareous substances which probably included bone, in his patent No. 649 in 1749. But in all the early porcelains, bone was combined with ingredients like soap-stone (steatite), lime, sand and often a glassy frit. It was Spode II (1754–1827) who used bone only with china clay and Cornish stone.[26] Bones

Table 1 **Ball Clays**

Cargoes from North Devon *a*

Port of Bideford

	Bristol	Liverpool	Chester	Gloucester
1691	7	1	1	2
1720	20	2	3	3
1735 ½ yr	14	3	3	2
1736 ½ yr	14	3	1	
1741 ½ yr	8	2		

Cargoes from South Devon *a*

Port of Exeter

to	Bristol	Liverpool	Chester	total tonnage
1736 ½ yr	2			
1741 ½ yr	1	2		
1750 ½ yr		3		400
1755		8	1	934
1758	1	10	1	1,465
1760 ½ yr	1	7		813
1765	1	20		2,381
1770	3	29		4,069
1776	9	18	1	2,484
1780	1	58		6,470
1785	10	77		9,995

Teignmouth

	total tonnage
1785	10,000
1811	17,000
1822	20,000
1832	15,000
1840	19,000
1852	30,000

Cargoes from Dorset *b*

Poole

	total tonnage
1760	10,000
1802	14,796
1808	22,000
1859	63,000
1861	75,000

Sources

a Bulley, J. A. (1955) The Beginnings of the Devonshire Ball Clay Trade. *Trans. Dev. Assn.* Vol. LXXXVII. Torquay.

b Hutchins (1861) *History and Antiquities of the County of Dorset.* London: Nicholls.

Table 2 **China clay and Cornish stone**

	Production of China Clay		Cornish stone
	from Cornwall tons *c*	from Devon tons *d*	from Cornwall tons *d*
1809	1,757		
1811	2,086		
1826	7,538		
1838	13,440		
1851	80,000		
1859	61,470		
1861	60,750	1,769	19,700
1865	97,750	8,570	25,500
1868	100,000	11,000	29,000
1874	150,500	33,309	42,500
1878	185,203	27,093	41,250

Sources

c Cock, D. (1880) *A Treatise, Technical and Practical, on the nature, production and uses of China Clay.* London.

d Howe, J. A. (1914) *Handbook to the Collection of Kaolin, China-Clay and China-Stone in the Museum of Practical Geology* (citing *Mineral Statistics of the United Kingdom*). London: HMSO.

Flint pickers on the beach at Criel Plage, Normandy c1890 (Courtesy Nora Walley)

of cattle only are used, the best being knuckle and shin of oxen. The bones were calcined in kilns similar to those for flint, but had wood staves instead of coal to prevent contamination of the colour by pyrites, etc. and to assist the draught: the bones, once ignited, burnt on their own to a temperature of about 1100°C. The subsequent processes were the same as those for flint.

This great advance added one more material which required to be ground for the pottery industry: but of the three basic raw materials for the preparation of pottery bodies – flint, Cornish stone and bone – the most interesting is the preparation of flint.

Sources of flint

In 1837 Simeon Shaw recorded that 100,000 tons of flint a year were supplied to the Potteries,[132] being obtained mainly from Gravesend, Brighton, the Isle of Wight and the south east coast of Ireland: besides these, there are deposits of flints on the shores of Lincolnshire and East Yorkshire,[156] and flints occur in great quantities on the shores of Norfolk, around Sheringham, having been carried there by the tide from outcrops which occur out at sea.

Flint is obtained today mainly from the chalk pits of southern England and from the coasts of Belgium and Northern France.[112] There are two kinds of flints used for pottery bodies, namely *chalk flints*, which are mined in the south east England chalk pits and have a crust of calcium carbonate, and *boulder flints*, which are chalk flints that have been washed by the sea and are mainly found on beaches.

Preparation of flint

Before being used in a pottery body, flints are calcined because this makes them easier to crush and grind, it burns off the very slight amount of carbonaceous matter but more particularly it disrupts the natural crystalline structure.[119] On heating, water is driven off to leave air spaces which lowers the specific gravity and assists in shattering the crystalline structure. Calcination is in brick-lined kilns which are filled alternatively with flints and coal slack in the proportion of about one ton of slack to 20 tons of flints. These materials are ignited at the bottom of the kiln and allowed to burn naturally until cool enough to handle. The calcined flints are then removed from the bottom of the kiln and are usually crushed before being passed to the grinding

The 'Albertine', of Newhaven, a *boulder* boat. These boats ran on to the open beaches of Sussex, loaded with boulder flints, & floated off by the high tide.
Sailed into Newhaven the flints were transhipped into brigs and schooners which plied in the trade to Runcorn. (Courtesy, M. Bouquet)

pans or mills.[37] Some flint watermills did not have flint calcining kilns, but at Cheddleton on the River Churnet, being favoured with the Caldon Canal running immediately beside it to provide transport, there were two such kilns; presumably the draught in the valley was sufficient not to require the hovels which more usually were built above the kilns. Although the calcination temperature and method was uncontrolled a temperature of about 900°C would be achieved and this was sufficient to render the flints friable and with a specific gravity of about 2.5.

Before the introduction of wet grinding, flint was crushed by stamp mills. These consisted of heavy timbers or iron shafts which were raised in succession and allowed to drop onto the stones or flints. When this process was used to produce the very fine flint, the dust would have been excessive, men dying from silicosis after only a few years working at this task. Benson's and Brindley's improvements removed this need, but stamp mills were still used for any intermediate crushing before the fine grinding on the pan:[70] the efficiency of the pan was increased and the grinding time reduced. In later years, stamps were replaced by jaw-crushers, or granulators:[147] an early reference to stamps of about 1793 is made in a memorandum from Josiah Wedgwood to James Watt 'Steam engine for the power of ten horses, to perform the following operations: . . . .

1 To grind flint in a pan feet diameter.
2 To grind enamel colours in a number of small pans . . . .
3 To work a stamper for pounding our broken saggars into a powder that will

pass through a coarse hair sieve – the floor of this stamper must consist of a large granite or boulder stone with a flat surface – it must be nearly square – three sides of plank the fourth to take off when it is to be emptied – this may be made in the manner of the stamping machines for stamping tin or other hard ores

4 To temper clay . . . ."[143]

Steam power

The pottery industry continued to expand towards the end of the eighteenth century and well into the nineteenth, and many of the individual manufacturers, wishing to have the milling operations under their own control, established grinding mills on their own premises. It appears that Josiah Spode I was the first master potter to install a fire engine[130] – the contemporary name for an atmospheric steam engine or pumping engine invented by Thomas Newcomen, and this was used to 'raise water to work a wheel'.[67] This engine is believed to have been installed as early as 1779 and was still in use 40 years later.[49] When Josiah Wedgwood was negotiating with James Watt for the purchase of a steam engine, a note giving details of Spode's engine was sent to Watt.[142] Spode's engine pumped water to a height of 13 feet 9 inches to flow over a waterwheel 12 feet diameter and 4 feet wide. This wheel drove a pan, 7 feet diameter, for grinding flint at the rate of 300 pecks per week, and some small pans for grinding enamel colours. A model of such an arrangement can be seen at Cheddleton; this is of Watt's 'Old Bess' built in 1777 for use at the Soho Works of Boulton & Watt in Birmingham.

Flint crushing by stamp mills

Rotative steam engines were not available until Watt patented his 'Sun and Planet' motion in 1781.[32] Wedgwood's engine at Etruria was ordered in 1782[145] and installed about 1784;[144] this was the first of Watt's engines to be installed in North Staffordshire; another was installed in 1793. A Watt pumping engine was installed in 1791 at the Fenton Park Colliery by Josiah Spode, Thomas Wolfe and partners, the lessees of the land and proprietors of that Company.[21] Later, in 1802, Spode installed a 10 hp rotative beam engine at his Works in Stoke, and replaced it with a Watt 36 hp engine in 1810[21] which was not removed until 1923. Beam engines were superseded either by individual electric motors, or, as at Mintons Works in Stoke, by a reciprocating steam engine, which has been preserved at Cheddleton.

Modern grinding

The method of grinding materials by pan remained in favour for many years after the introduction of the ball mill, or cylinder grinding mill, by J. R. Alsing.[164] This method of grinding is now so firmly established that the use of pan mills for grinding pottery materials is almost extinct; where it does survive it is confined usually to the grinding of bone.

Great improvements in grinding methods have taken place in the past twenty-five years; the continuous grinding in Hardinge mills is widely used, while vibro-energy mills are being used to an increasing degree.

Left: Boulton and Watt Rotative Engine used at Etruria, 1801–1912 (courtesy of Messrs. Josiah Wedgwood & Sons Ltd).[142]

Right: Drawing of Watt's engine for Mr Wedgwood 1784 (Boulton & Watt Collection).

Below: Watt's single-acting engine for pumping water for draining mines, 1788. The boiler is placed in an outhouse, and the steam passes to the cylinder E, which is kept hot all the time by a separate steam jacket. F is the separate condenser and H. an air pump. (after J. Farey 'A Treatise on the Steam Engine').

Below right: Watt's double acting rotative steam engine of 1784. The parallel motion bars are seen on the raised end of the beam. Note also sun-and-planet gear (after J. Farey ibid).

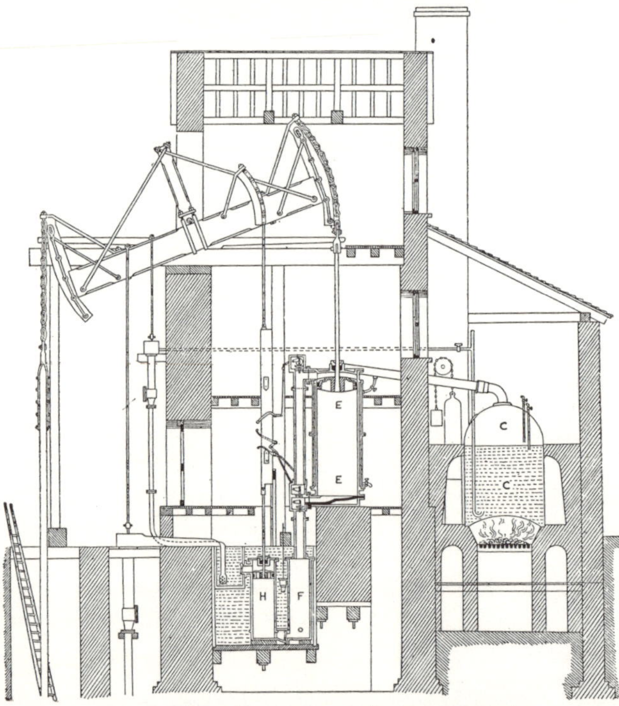

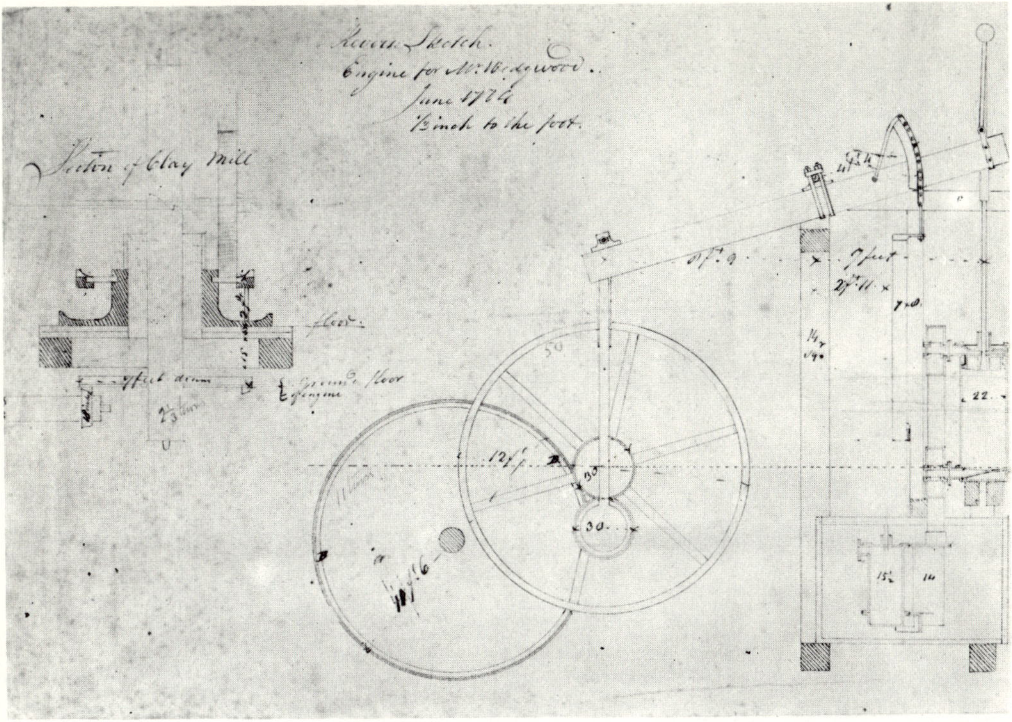
Engine for Mr Wedgwood
Section of Clay Mill
floor

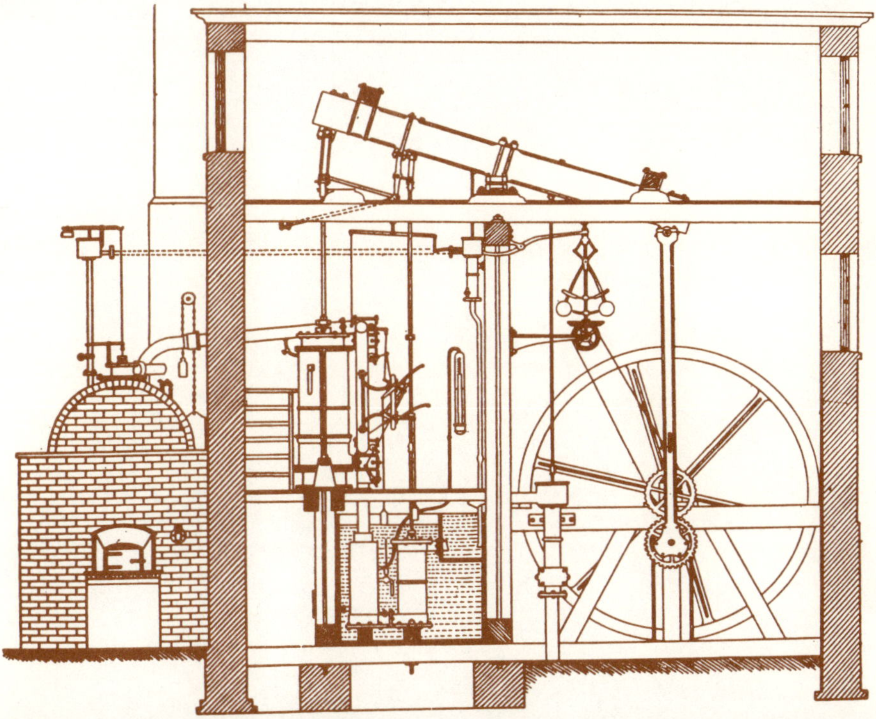

Cheddleton Flint Mill *above* after restoration, February 1972, *left:* c.1894

The Cheddleton Flint Mill

History

The history of the Mill is still obscure. The earliest record found so far shows that in 1253, in the reign of King Henry III, there was a Mill at Cheddleton in possession of one Roger at the Mill.[55] More recent, is the will of Josiah Ffiney, dated 1694, in which it states that he 'owned all those water corne mills comanly called Chedulton Mills . . .'.[36] Cheddleton was an ideal site for a flint mill because, not only was there a plentiful flow of water along the River Churnet, a tributary of the River Dove, but also a main turnpike road passed by. This road, from Sandon to Leek, was the direct road from Lichfield to Manchester and was crossed by the important staging road from Derby to Newcastle and Stoke, both roads having been turnpiked before 1763. The bridge at Cheddleton was changed from one suitable only for horses into a cart-bridge in 1687, and was widened in 1784.[141] By this date, the Caldon Canal was opened, and the Mill, with a wharf, is shown on the map published in 1795 by Aikin, which indicated the routes of the Trent and Mersey (Grand Trunk) Canal with the branch canal from Etruria to Froghall.[3] This branch, the Caldon Canal, was cut in order to extend the markets of the limestone quarries at Caldon Low (16 Geo. III 32) and was completed, at the same time as the Grand Trunk Canal, in 1777.

Lying between the canal and the River Churnet is the Ffiney Mede, an area of meadow with a raised embankment along the river course: in times of severe flooding, the river water would flow over into this meadow which slowed the force of the torrent and so protected the waterwheels from damage. Aikin's map also shows two Flint Mills at the south of Stoke on the River Trent and one north west of Froghall, at Consall.

James Brindley (1716–1772) supervised the surveying of the Caldon Canal but died on 27th September 1772 before construction began.[138] It is possible that Brindley also constructed the North Mill at Cheddleton for grinding flint:[17] the upper walls (from the grinding floor level) were rebuilt sometime after 1783, the year in which the brick tax was introduced and when the size of bricks was increased.[28] The construction of the North Mill indicates that it was for the grinding of flint rather than the lighter material of corn. Brindley was skilled in the engineering of flint mills, having begun as early as 1748 by erecting a windmill for John and Thomas Wedgwood for this purpose on an area of Burslem called The Jenkins.[74]

Facing page: The flint kilns. *Above*, the stone retaining walls & kiln mouths, with waggon (dated 1803) on cast iron plateway, *Below*, the top of the kilns, calcined flints in the nearer kiln.

An early document which relates to flint mills is one dated 24th April 1744, in the Stoke-on-Trent City Museum: this is an Agreement between Thomas Adams, Stephen Stringer and Thomas, for John Wedgwood (sic.) 'to take, farme and rent ye old ffurnace at Lawton (Church Lawton) and convert it into a fflint mill . . . ." In Thomas Whieldon's Account and Memorandum Book, also in Stoke-on-Trent City Museum, reference is made to his flint mill in the entry of 3rd December 1760. Whilst the Ivy Mill in the Moddershall Valley was built by Thomas Benson and John Gallimore with Mr Astbury's son providing the capital; this mill would have been built even earlier than the ones mentioned.[50]

According to Smiles' Life of James Brindley an entry in his pocket book for 15th March 1757, gives details of a large water-driven flint mill to be erected for Thomas Baddeley at a place called Machins' of the Mill, near Tunstall;[77] this mill seems to have used stamping machinery for pounding and grinding the flints. John Baddeley had interests in other flint mills from which he derived a fair income by selling the ground materials to other potters.

Benson's 1732 patent expired in 1746, and one must assume that these patents were well known to Brindley in 1757, for Smiles attributes the invention of an improved mill for grinding (as that at Cheddleton) to Brindley. The South Mill at Cheddleton is older and was the original corn mill, later reconstructed to grind flint. It is not known for certain when this alteration took place but it seems a number of enlargements occurred one of the most important taking place sometime after 1783 (judging from the size of the bricks).

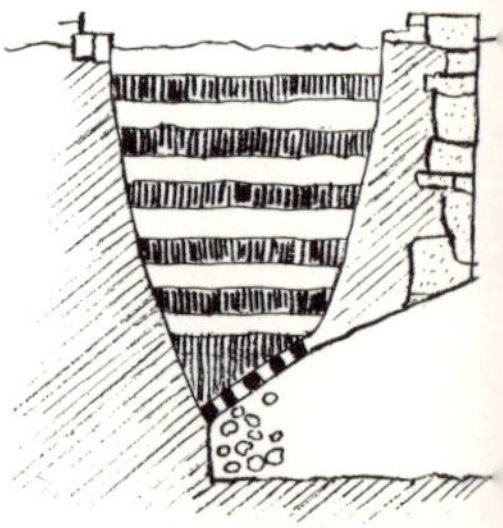

Section of flint kiln showing alternate layers of flint and coal.

Advertisements for the sale by auction of Capital Flint Mills (Lots 2 and 3) appeared in the Staffordshire Advertiser of February 18th and 25th 1815: these show that both mills were grinding flint at that time. The design of the mechanism in the South Mill bears a strong resemblance to that in the North. It is possible, of course, that the Cheddleton Mill was grinding flint in very small quantities using the original corn grinding stones long before the North Mill was introduced: Nicholls mentions that 'Aaron Wood, younger son of Ralph Wood (1676–1753) Miller of Cheddleton and Burslem, was apprenticed to Thomas Wedgwood in 1731. He was the best modeller of the day and modelled for all the manufacturers.'[93] With Ralph Wood's very important connections with the potters of the day he may have undertaken the grinding of flint using stone similar to the one displayed at Cheddleton, made of Mole Cop stone.

A circular structure, believed to be a sunpan, recently excavated is of bricks made after 1783, and this disclosed several pieces of pottery made during the last part of the nineteenth century; this suggests that it may have been filled in before 1900. In the photograph of the mills c 1894, there is no chimney to

he London Road
e Potteries, about
s, and are in good
as been recently
me and place will
piece of meadow

nad of the above
ase money may

out reserve.

ION.
SON,
Market Drayton,
of March, 1815;
and fat Sheep.
Castle aforesaid,
ng; consisting of
ful calving heifers,
one hundred fat
y be agreed upon

in the choice of
y comment;—the
ne to early profit,
and the whole in
ncers beg leave to
eleven o'clock in
otice.

Nc.

CTION,
SON,
hursday, 22nd and

Husbandry, and part
g to the late Mr. T.
esball, in the county
ry cows in-calf, 11
alves, one good bull,
e draught mare, five
ling, four years old,
ht mare, 3 two-year
ing, one ditto of the
dung cart, long cart
h, 2 wheel ploughs, 2
set or pair of twins,
bridles, 3 ladders, 2
ls, and about half a

bedsteads and hang-
al sets of mahogany,
other chairs, with
ining, dressing, and
wo linen chests, pier
nd drawers, a good
hen grate, ash grate,
bottle jack, fenders,
nned plate warmer,
pails, gauns, tubs,
d brewing utensils,
and numerous other

ning at 10 o'clock.—
, will be sold on the
evening a part of the

CION,
SON,
Haven and Bell Inn,
on Saturday, 25th

---

half-past ten o'clock.

Capital Flint Mills, Brewery, and Lands.

TO BE SOLD BY AUCTION,

IN LOTS,

At the house of Mr. Ridout, the George Inn, in Leek, in the county of Stafford, on Thursday, the 2nd day of March, 1815, between the hours of four and six o'clock in the afternoon;

LOT 1.

ALL that Messuage or Dwelling-house, with the several Buildings thereunto belonging, now used as a Brewery, and in the occupation of Messrs. Joul, situate at Cheddleton, in the said county.

The House is a good residence, and now occupied by Mr. Smith, with a large Yard and Garden, and the Buildings consist of Brewery-vaults, Store-room, Coopers' shop, and two Stables with stalls for eight horses. Cowhouse for eight cows, Piggery, &c.

Also all that Malthouse adjoining, capable of making 4,000 bushels of malt per year.

For supplying the premises with water, there is a large reservoir 36 ft. by 6 ft. fed by a constant spring of excellent water through lead pipes.

The above premises are subject to an agreement for a lease of which there will be five years to come at Lady-day next.

Also all that Messuage or Dwelling-house adjoining the aforesaid premises, as now in the occupation of Mrs. Smith, as tenant at will.

Also all that other Building adjoining the said premises, used as a warehouse, with the chamber over the same, 36 ft. by 13 ft. as now in the occupation of Mr. Robert Coopers, as tenant at will, and which might at an easy expence, be converted into two Dwellings.

LOT 2.

All those two Flint Mills, with Arks, Machinery, Wash Tubs, Pans, two Flint Kilns, Lime Kiln, two Stables, for five horses, Store Room for tubs, large Yard and other conveniences, now in constant work, capable of grinding 50,000 pecks per annum.

And also all that piece of excellent Land adjoining the said Mills, called the Mill Hulme, containing about an acre, possession of which with the Mills, may be had on payment of the purchase money.

LOT 3.

All that piece of Meadow Land adjoining the said premises, containing near two acres.

The whole of the above premises are in a ring fence, situate on the river Churnet, adjoining the banks of the Caldon Branch of the Grand Trunk Canal. The Mills have a regular and constant supply of water, and afford such an opportunity as seldom happens for establishing any large manufactory, where facility of turning and water carriage is required.

The whole is exonerated from Land Tax, and Tithe Free.

Cheddleton is situate about 3 miles from Leek; 6 from the Staffordshire Potteries; 7 from Cheadle; and about 13 from Stone; and through which there is a good turnpike road to the above places, as well as to Sandon.

John Hammond, the Miller, will shew the premises; and for further particulars apply to Mr. Weston, Rudyerd Hall, near Leek; or at the office of Mr. CRUSO, in **Leek.**

Leek, 8th Feb. 1815.

To be **SOLD** by PRIVATE CONTRACT,

S, 9, 28, at the Rose and Crow
torney, Mr. Brown, Dobcross, S
S. Coates, Sunderland, grocer,
cise Tavern, Bishop Wearmout
Sunderland.

**J. Wisbey, Thaxted, Essex, jo
28, at the Red Lion, Petty Curr**
ney, Mr. Fisk, Cambridge.

E. and S. Wileman, West Hout
21, 28, March 28, at Guildhall.
and Haddon, Angel-court, Throg

J. Shaw, Plymouth-dock, iron
28, at the Tontine Inn, Sheffield.
and Brown, Sheffield.

W. Greenwood, Elwick, York
March 28, at the King's Head In
Barrett, jun. Bingley.

T. Smith, Old Barge House,
21, 28, March 28, at Guildhall.
Searle-street, Lincoln's Inn.

F. Mullet, St. Mary Axe, me
28, at Guildhall. Attornies, Me
ples, and Pearce, Frederick's p

J. Harries, St. Thomas the A
Feb. 28, at the Old London Inn
Brutton, and Ford, Exeter.

BANKRUPTCY

**William Macmichael, Bridgn
London, carpet manufacturer,
Castle Inn, Bridgnorth.**

BANKRUPTCIES

**Richard Fillis, Plymouth,
Great Yarmouth, Norfolk, boo**

DIVIDENDS AT

March 7, T. Sedgwick and
merchants. Feb. 28, F. Thom
jun. Paternoster row, brokers.
G. H. Eades, Great Eastcheap,
kins and W. Cowper, Lincoln's
25, S. G. Mesle, Finsbury plac
27, D. Fowler and A. Anstie, G
Feb. 18, W. Sharp, Threadnee

DIVIDENDS IN

Feb. 28, J. Owen, Oswestry,
Inn, Chester. March 2, W. B
dealer, Crown Inn, Stafford.
Freeth, T. Cartwright, and A.
ford, brewers, White Hart In
8, Messrs. Welby and Bellairs,

PARTNERSHIP

H. Duckworth, jun. R. Hebg
pool. W. Weetch, J. Allowa
woollen drapers, tailors, and
Buttenshaw, Holborn Bridge, g
Finch lane, St. Michael, Corn
son, Newark-upon Trent, and
tingham, millers. J. Blankero
Bancroft, and E. Bancroft, Ch
ders and S. Wilson, Newgate
Papillon, P. F. Papillon, and
mondsey, Surrey, calico print
Heaton and J. Lancaster, Bolt
masons. T. Underhill, jun.
brass founders. S. Appleton an
Hull, coal dealers. D. Ross a
street, merchants. J. Hibber
lard's court, Crutched Friars,
chants. W. Deakin and G. I
near Sheffield, forgers and ste
Birch, High street, Whitech
sen. and B. Hickson, jun. Ki
ship and insurance brokers. an

Left: From the *Staffordshire Advertiser* of 18th February 1815.

the big slip kiln but there is no obvious explanation for this. The second and smaller slip-kiln was built about 1953 for the drying of titanate.

The Flint Mill was one of three businesses within a small industrial complex: the others being an ale and porter brewery, and a silk mill. The brewery is mentioned in detail in the 1815 advertisement and at the time was in the occupation of Messrs Joule (see opposite). In 1860 the fermenting rooms contained stone and wooden squares, and a steam engine provided power. Opposite the brewery and flanking the towpath of the Caldon Canal was the silk mill of three storeys with a brick chimney stack. The ground floor being built of stone. It seems that it was in use as a silk factory as early as 1838 in its present form when it contained 180 dozen spindles for throwing silk and 7 looms all worked by a 6 hp steam engine. In 1839 the silk industry was in a declining state and by 1855 the factory was unoccupied. It has been used as a warehouse in recent years.[135] On the other side of the Sandon – Leek Road, paper-mills have been established since about 1797: in that year William Adams bought 10 acres of a field known as Butchers Meadow near the River Churnet.[94] Here he built paper-mills principally for making paper for use at his factories in the transfer printing department and he also sold to other manufacturers.

These mills were later let to the Fourdrinier family and were subsequently purchased by Messrs Brittains Limited.

Flint mill process

At Cheddleton the calcined flints were loaded from the flint kiln bottom onto trucks which were pushed along the plate-way to a point below the trap-door in the first floor of the mill where they were raised by a hoist, and the flints subsequently charged onto the pan. The grinding pan consists of a floor paved with chert stones, mined usually in North Wales near Gronant in Flintshire and also in Richmond, Yorkshire. This stone has a very high silica content and is similar to an amorphous form of flint. The thickness of the paved bottom is between 9 and 12 inches and the stones, called 'pavers' are quarried in such a way as to make a consistent thickness of pavement, which in effect is the lower mill stone of a conventional grist mill. The function of the upper mill stone is reproduced by pushing heavy stones over the surface of the pavement by means of arms attached to the centre shaft, which is driven by either the waterwheel or another source of power. In a large pan there are usually four cast iron sweep arms to which are fixed thick wooden hanging arms to protect the iron arms from abrasion caused by the heavy runners that are being pushed round. The runners are stones of a high silica content,[43] also called chert, but more heavily veined with limestone and therefore softer; these are quarried in Derbyshire near Bakewell.[40] In Cornwall, the grinding of Cornish stone was in similar pans where the staffing of the pan, the pavers and runners were all of the same stone that was being ground.[51] The runners wear out more rapidly than the pavers but are more easily replaced. Indeed they should be replaced or repositioned fairly frequently in order to avoid the difficulty of sucking whereby they will run in their own grooves and not contribute to the grinding action for which they are intended. Calcined bone, being much softer, does not create such grooves and it is easier to keep a pan for bone for a very much longer time before replacing the grinding stones.

When it is considered that the material has been ground finely enough a wooden bung or plug is removed from a hole in the pan side allowing the ground material, which is of a creamy consistency, to run off into the wash tub. This is a circular vat with a central shaft to which are fixed wooden gates with long handles and these gates are rotated by hand.

A considerable amount of water is added to the ground material and thoroughly agitated to ensure that every individual particle of ground material is properly suspended in water. The wash-tub is, therefore, a simple form of classifier: the insufficiently ground particles drop to the bottom, while the overground *fines* float on the surface. When this is achieved the contents of the wash-tub are run off into the settling ark, any coarse material passing to the arks will settle first at the bottom, and that which is sufficiently ground will settle next, leaving the very finest material which has perhaps been ground too fine to be the last to settle. This settling ark, or tank, is provided with a series of holes in its side (plug-plank) through which the water is drawn off as the ground material settles. When the water has been drawn off through this plug-plank to the level where some of the ground material is also flowing away, the bung or plug is replaced and the contents of the ark are then pumped onto the drying kiln floor where it remains until most of the moisture has been evaporated.

When sunpans were in use, about 4 inches of slop flint were drawn off into a pit of the sunpan where the water evaporated.[154] When dry, another layer of slop material was added, and this process continued until the dry material was about 18 inches thick. Before complete drying has taken place the stiffening mass is cut up partly to hasten drying but more particularly to make it easier to remove the dried material in convenient blocks (see page 50). Flint which is considered to be insufficiently ground is dug out of the settling ark and separated into two grades. The coarser is termed *bitstone*,[73] being about one eighth of an inch across and used for putting in saggars in which ware for glost firing is placed, to prevent the sticking of the ware to the bottom of the saggar; the finer material was originally used for placing bone china (bedding) ready for the biscuit fire, but this began to be discontinued in about 1932 when calcined alumina was substituted because of the very high incidence of silicosis caused by the use of flint.[102] The use of flint for china bedding was forbidden in 1947.

The milling and preparation of stone and bone followed the same procedure, except that stone was not calcined.

Facing page: Two examples of entries in the account book of Robert Thornhill,[40] in which are records of deliveries of chert to Cheddleton. Those charged to Thomas Cheadle cease after 1800, and from 1802–1810 are charged to William Adams.

Preparation of glazes and colours

Glazes

The preparation of fluid glaze for dipping the ware involves the following operation: weighing out, mixing, grinding, agitation and the adjustment to the correct pint weight and viscosity for application.

In the olden days the grinding of glazes was on pan-mills similar to stone-grinding, except that the staffing was generally of wood, to minimise iron

10 Mr Thomas Cheadle Dr — Out Lanes Stafford

Date	T	Cwt		£	s	d
1796 August 23	13	16	Delivd at Cheadleton at 1.12.6	22	8	6
1798 April 18	9	12 2	at Do ... Do	15	12	9
1799 Octr 7	4	10	at Do ... Do	7	6	3
1800 Augt 11	9	4	at Do ... 1.15.0	16	2	..
1802 Jany 2	6	..	at Leek ... 1.12.6	9	15	.
~~Dec 21~~	~~3~~	..	~~at Do ... Do~~	~~4~~	~~17~~	~~6~~
1803 May 20	By a Ballance then due			10	14	6

Date	
1797 May 23	By a Bill on Acct
1799 Nov 14	By a pocket of Hops
1803 April 27	By a Bill on Acct
1805 Nov 16	By Bill of Cr

1 Mr William Adams Dr — Cobridge Staffordshire

Date	T	Cwt		£	s	d
1802 July	9	8	Delivd at Cheadleton at 1.15	16	9	
1803 Nov 20	17	..	Delivd at Cheadleton at 1.15	29	15	.
1804 May 28	6	..	Delivd at Cheadleton at 1.15	10	10	
Decr 7	5	..	Delivd at Cheadleton at 1.15	8	15	
1806 Jany 2	10	..	Delivd at Do at 1.15.0	17	10	
Decr 26	10		Delivd at Do at 1.15.0	17	10	
1808 Jany 20	10	5	Delivd at Do at Do	17	18	9
1809 May 1	18	18.1	Do at Do — Do	33	1	11
1810 March 8	16	1.1	Do at Leek Do —	32	2	6

Date	
1803 May 21	By a Bill
1804 May 11	By a Bill
1805 April 24	By a Draft
1806 April 29	By a Draft
1807 April 28	By a Bill
1808 May 9	By Bill
1809 July 11	By Bills ...
1810 June 8	By Bills

102.12

Cheddleton Mills and Ffiney mede. The larger map (scale 25 inches to 1 mile – based upon the Ordnance Survey map of 1880) shows the important site between the river, which provided power, and the canal and turnpike road. The inset plan (scale 1 inch to 40 feet) shows the original layout of the mill c1785 with later additions. An extension *(a)* was removed c1953: it accommodated a wash tub.

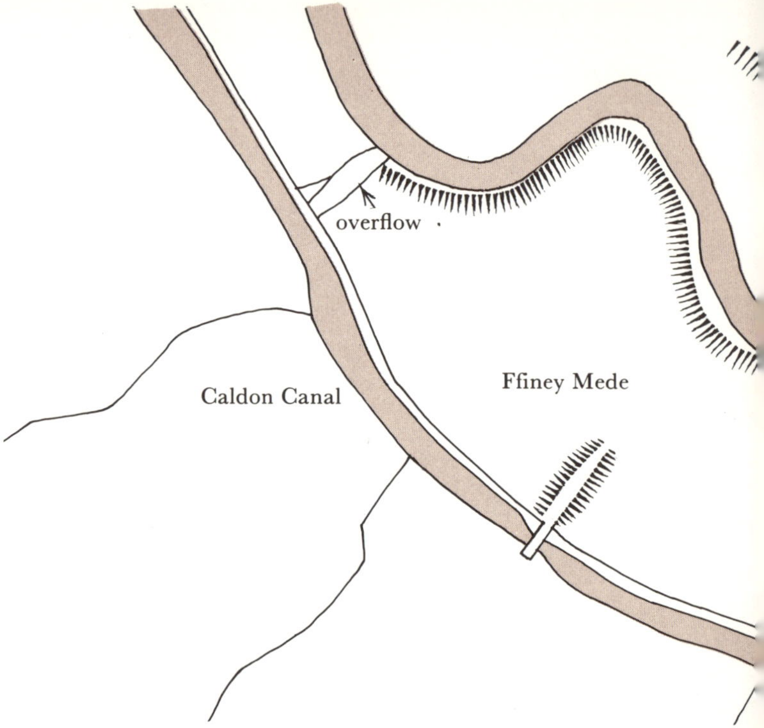

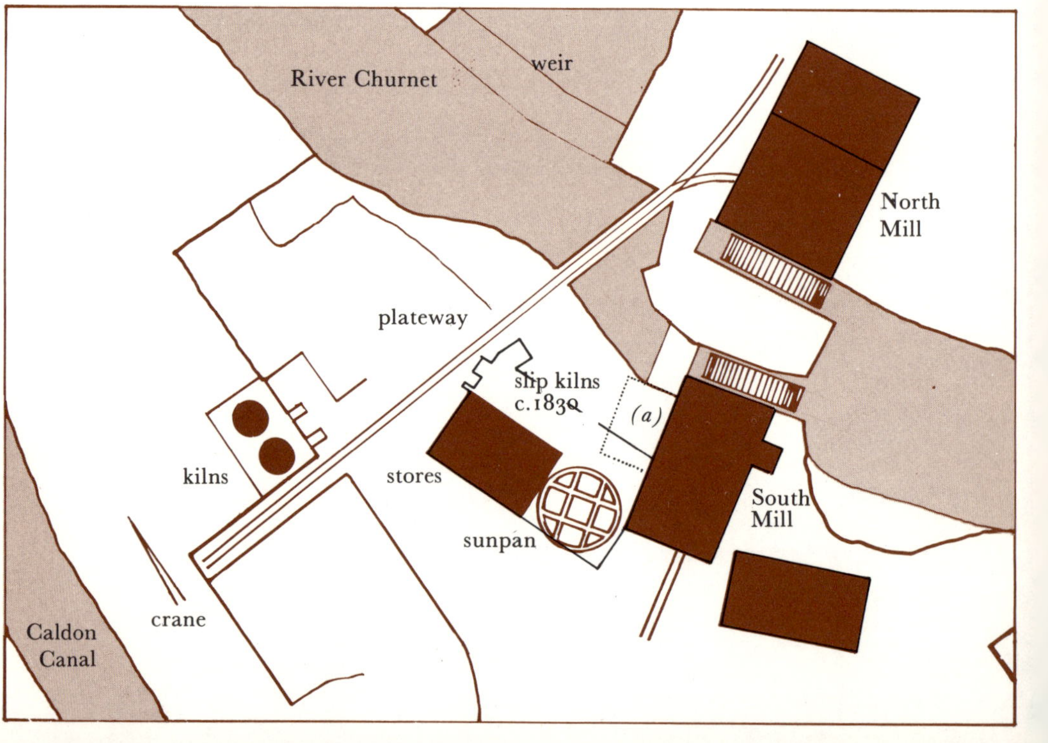

N
River Churnet
towing path
to Leek
weir
Brewery
stables
winding
hole
Silk Mill
to Stone and Sandon

1
2
3
6
4
8
5
7
10
9a
11
9c
14
13
12

1 **The Caldon Canal** was constructed to carry the Limestone from the quarries at Caldon Low; it was surveyed by James Brindley in 1772 and completed in 1777, the same year in which the Grand Trunk (Trent and Mersey) Canal was opened.

2 **Narrow boats** brought the stones, flints and coal to Cheddleton Mill and returned with the ground flint to the Potteries. The flints came from South East England, East Anglia and France by ship to Runcorn, thence by narrow boat via the River Weaver, and the Trent and Mersey Canal.

3 **The Crane** was used to unload the stones from the narrow boats and lower them to the storage pavement below – later to load them onto the waggon.

4 **Pavers,** the chert stones from North Wales, Derbyshire or Yorkshire used to pave the floor of the grinding pan, which may be from 10″–14″ thick. This represents the bedstone of a conventional corn mill.

5 **Runners** (runner stones) are of chert from Derbyshire, North Wales or Yorkshire, with heavier veins of limestone and softer than the chert used for the pavers.

6 **Flint kilns** for calcining the flint. There are two kilns at Cheddleton: there are no hovels, the natural draught being sufficient to keep the fires burning. The kilns were loaded with alternate layers of coal and flint and burnt for about three days. One hundredweight of coal would be used for one ton of flints.

7 **Calcined flints** are drawn from the bottom of the kiln and taken to the first floor of the mill.

8 **The Plateway and waggon:** the plateway, discovered buried in the flint kilns, was probably laid down about 1777 when the canal was opened. (Cast iron rails were first introduced at Coalbrookdale in 1767). The waggon has '1803' carved on the axle tree and was brought from the Spode Works. The distance between rails is 2 ft 1 in.

9 **Auxiliary shafting** driven by the vertical shaft of the pan drives the **hoist,** or **'teagle'** (9A) which raises the flints to the first floor, the **water pump,** which provides water for the grinding, and the **slip pump,** (9C) which pumps the correctly ground material from the settling ark over onto the slip drying kiln.

10 **The pan** would take a 'charge' of about 1½ tons of calcined flints and would grind for 24 hours. Then the thick creamy fluid would be run off into the wash tub. The grinding weight would be about 32/34 ounces of ground flint in a pint.

11 **The Washtub:** in order to ensure that every particle of flint is completely dispersed, the flint is mixed with a large volume of water and, by turning the wooden gates, the whole is thoroughly agitated. The main purpose is to separate the coarse, insufficiently ground particles from the fine material. The pint weight at this stage will be about 22 oz/pint. The 'fines' thus dispersed are discharged to the settling arks, while the coarse washings are returned to the pan again. This coarse fraction is called 'knockings', or 'knottings'.

12 **The settling ark,** into which the flint, suspended in water, is run off to settle.

13 **The plug plank.** As the flint settles, the clear water is drawn off by removing the plugs in the plug plank, until the level is reached when the flint itself appears. The plug is replaced and the slip pump started to pump the slop flint to the drying kiln or into barrels if it is required in 'slop' form. The pint weight of the flint has now been increased back to 32/34 oz/pt.

14 **The Slip drying kiln** consists of a flat pavement of firebricks, with low side walls of brick, over flues through which the flames pass. The water in the slop flint is evaporated and, when all of it has been driven off, the flint is removed in blocks, or 'cake'.

The grinding pan

Plan

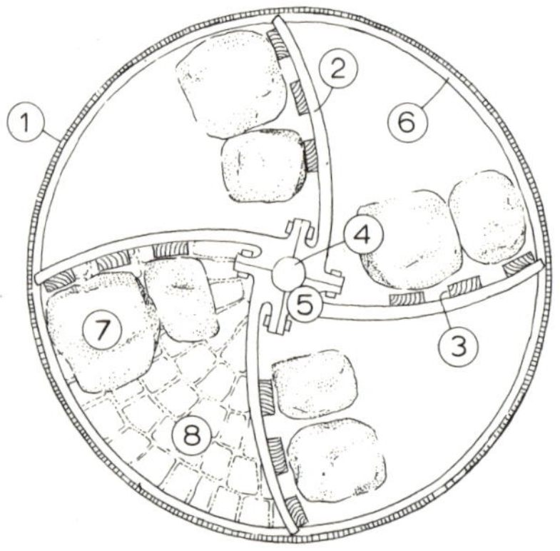

Elevation

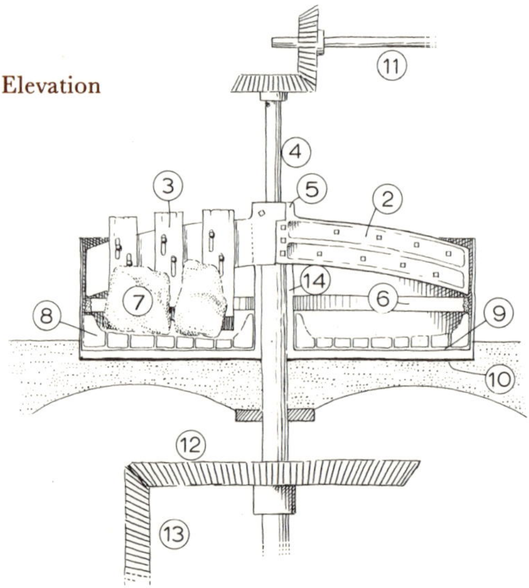

1 **Staffing,** of wood, or iron or steel plates: also of stone, either as flagstone, or masonry (especially in Cornwall).

2 **Sweep arms** of cast iron: they can run free on the shaft, being tightened by a brake strap or drum.

3 **Hanging arms** of oak, with slots to enable them to be adjusted as the pan bottom wears. These arms are fitted with iron 'toe plates'.

4 **Shaft,** driven by the 'wallower', and extended to provide additional drive for auxiliary mechanisms, like the hoist, water pumps and slip pumps. The speed may be 12–18 revolutions per minute, depending on the diameter.

5 **Brake drum,** to provide a slip mechanism in case a stone becomes jammed and is likely to cause a breakage. The brake band contacts the driving lug on the sweep arms.

6 **Slug iron,** a continuous band of iron to ensure that the runner stones do not fall against the staffing and that the sweep arms are kept at their correct respective distances.

7 **Runners,** or **runner stones,** of Derbyshire chert, may vary from 1–30 cwt.

8 **Pavers,** or **paving stones,** of Welsh or Yorkshire chert. These pavers form an even bottom, which might vary in thickness from 10 in to 14 in, or even to 16 in. The pavers are grouted in with broken flints and broken china pitchers. The pan bottom shown here is worn.

9 **Levelling dirt,** or **'pricking' dirt,** on which the pavers are set.

10 **Clay seal,** to ensure that the water does not percolate. This is laid on a cast iron plate, or directly onto the masonry over the structural arch.

11 **Horizontal shafting** for driving the hoist and pumps.

12 **Wallower**

13 **Pit wheel**

14 **Centre tub** of cast iron allows the shaft to pass through from below, yet ensures that the 'charge' of flint and water is retained.

Cheddleton Flint Mill

A Museum of Industrial Archaeology
at Cheddleton near Stoke-on-Trent

Cheddleton
Flint Mill
A Museum of
Industrial Archaeology
at Cheddleton near
Stoke-on-Trent

Area map

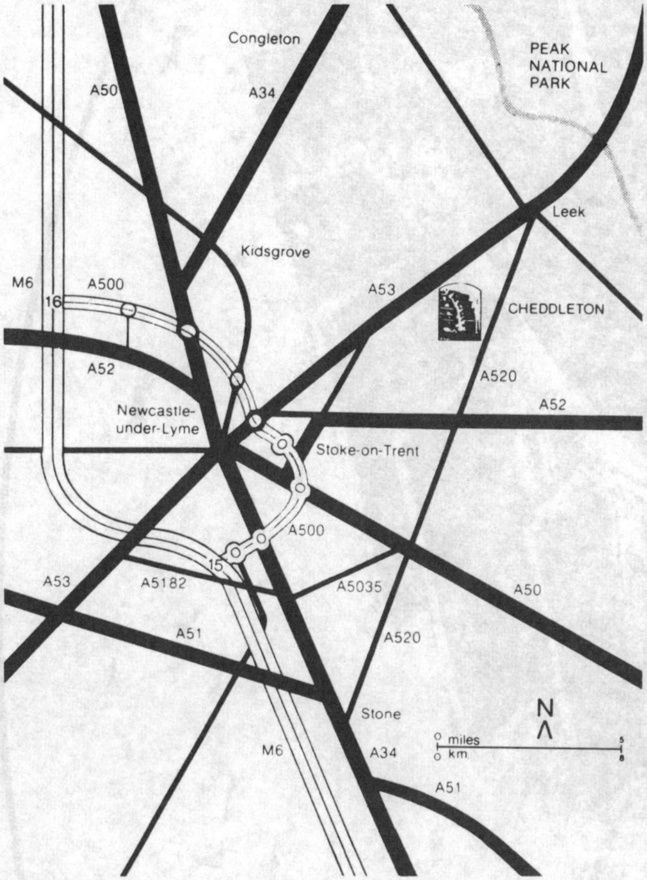

How to get there

Cheddleton Flint Mill is at the bottom of the valley, to the West side of the A520. When approaching by road from the *South*, look for a 'Cheddleton Flint Mill' sign on the left hand side immediately before the bridge. Approaching from the *North* on the A520, cross the bridge and look for the sign on the right hand side. Access is by a narrow lane which is unsuitable for heavy vehicles.

General Information

There has been a mill standing at Cheddleton on the River Churnet since the 13th century. Originally the South Mill was used to grind corn. During the Industrial Revolution the growing ceramic industry in Stoke-on-Trent created a demand for powdered flint, which is used to improve the texture of earthenware and tiles. The South Mill was converted to grind flint and the North Mill was built in about 1800. The Caldon Canal links the Mill to the Potteries for transporting materials and ware by narrowboat. The Mill has now been preserved as a Museum of Industrial Archaeology by the Cheddleton Flint Mill Industrial Heritage Trust.

Admission is free.

Outdoor exhibition

You are welcome to visit at any time during daylight hours. The museum has captions outside to explain each exhibit and how the Mill worked.

Building interiors

The buildings are open
1.00 until 5.30 pm throughout the year.

Parking

The Mill has a small car park.

For daytime visits please write to:

Stoke-on-Trent

For visits please write to:
Mr E Royle, 5 Caroline Crescent, Brown Edge,
Stoke-on-Trent

A small pulverising cylinder, or ball mill. Manufactured by William Boulton under licence from J. R. Alsing. A No. 7 size mill of this type is in the Cheddleton collection

contamination; nowadays, ball mills lined with porcelain or hard rubber have completely replaced grinding pans. It is best to grind the harder materials with water first, so that the risk of overgrinding the softer ones is avoided: all the materials when mixed, are ground together in water, the pint weight of the fluid on the mill being 36 ounces.

Many glaze compositions, and also porcelain and parian bodies, comprise a mixture of materials which have been fritted before being mixed with the raw materials. Fritting is necessary to convert water soluble ingredients, like borax, into insoluble compounds, usually silicates: this is necessary to ensure that a homogeneous glaze is produced, and that most of the gas producing substances have settled down before the glost firing. Lead compounds are fritted to ensure that no soluble lead is present which could be dissolved in the gastric juices of the human body.

Frits are generally prepared in intermittent kilns; these may be of the rotary type, in which the charge is heated in a refractory-lined steel cylinder by an oil burner, or in a frit kiln of the reverberatory type by long flames from a coal fire placed at one end with the chimney at the other. When the charge has been melted completely, it is poured into water which shatters the frit thus assisting subsequent grinding.

Colours

The preparation of colours for the pottery industry is a highly specialised science concentrated in the hands of a few colour manufacturers. But years ago, many pottery manufacturers made their own colours, whilst there were a large number of specialist small colour makers, often building reputations for themselves by preparing certain groups of colours, e.g. crimsons and purples derived from gold.

Unlike many colours, ceramic colours can only be derived from a narrow range of inorganic metallic oxides that do not decompose at firing temperatures of 700°C or more. Mixtures of these oxides with fluxing ingredients can provide sufficient tones throughout the whole spectrum.

It is the preparation of these mixtures which varies greatly and on which depends the success of the colour.

Recipes for colours consist of two parts: a list of ingredients with proportions, and the method of preparation. Published recipes usually omit the all-important second part dealing with preparation. Nearly all colour recipes contain ingredients which form gases on being heated and it is necessary to ensure this does not happen when the colour is on the ware during the decoration fire. Colour mixtures are calcined before grinding to drive off any gas and to render the mixture more stable and homogeneous in use. The process can include most of the following stages: weighing out the ingredients, thorough mixing by hand followed by sieving the dry materials several times to break up any lumps and to assist in mixing, calcining this mixture in saggars, crushing the resultant calcined mix, repeated washings to remove soluble salts, fine grinding in water and finally drying. More than one calcination may be necessary and perhaps each calcination may be at different temperatures: usually the colour is calcined to higher temperatures than it is likely to experience in subsequent decoration firings to avoid the

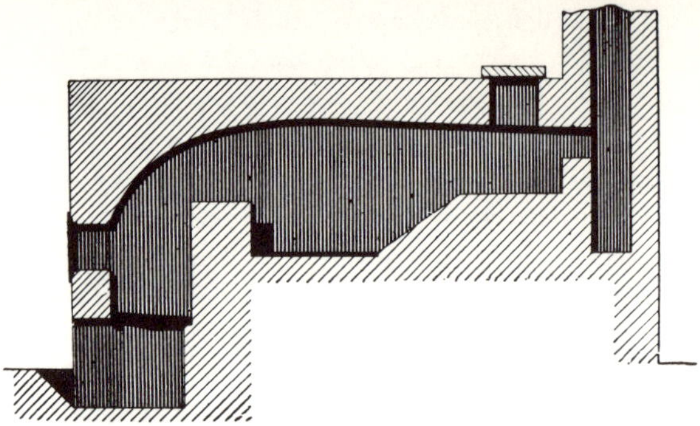

Frit Kiln of the reverberatory type The powdered materials are introduced into the kiln through the opening in the roof. The firebox is at the left side: the molten frit is *run off* from the lowest point of the centre of the hearth.

risk of any changes occurring when it is used on the ware. There are few exceptions to this, and such colours have to be used with great care to avoid decomposition and change of colour.

An example of a recipe for an earthenware colour for printing underglaze will illustrate a typical process:

BLACK

First Process

4 parts Black oxide of Cobalt

4 parts Crocus Martis

3 parts Green oxide of Chromium

Method Calcine in China biscuit oven in forebung just below bag (i e about 1350°C), grind, and wash well until water is clear.

Second Process

3 parts of first process mixture

2 parts ground Chromate of Iron

1 part Red Lead

Method Calcine in saggars in China biscuit Oven in similar place as before. Grind, wash very well and dry.

The following list shows the principal metals used in the preparation of ceramic colours: but the combinations in which they are associated varies greatly according to the uses for which they are intended.

BLACK Nickel; Iron; Manganese; Chromium; Cobalt

BLUE Cobalt

MULBERRY Cobalt with Manganese

TURQUOISE Tin and Copper

GREEN Chromium and Zinc

PINK Chromium with Tin; Gold with Tin

CRIMSON, MAROON Gold with Tin

PURPLE Gold with Tin

BROWN Iron; Zinc; Chromium

RED Iron; Selenium

YELLOW Antimony; Lead; Iron; Vanadium with Tin; Uranium

More information on glaze and colour preparation may be found in the books covered by Reference numbers 18, 39, 41, 60, 120 as well as specialised books on the modern technology of pottery.

The North Mill

The Mill building would have been built sometime before 1783 as the upper walls from grinding floor level of this Mill were rebuilt with the new sized bricks.[28] The waterwheel is 22 feet in diameter and 5 feet $9\frac{1}{2}$ inches wide. There are two sets of ten oak arms fastened on to ten cast iron hubs; these secure the spokes to the shaft. The outer rim of the wheel is made up of cast iron segments with boxes for the starts to fit in. There are 40 L-shaped floats each 14 inches deep made of larch-wood. The shaft is of cruciform section measuring $15\frac{1}{4}$ inches across the flats (it is considered that this shaft is later than the one in the South Mill). The wheel is controlled by a sluice, or *shut*, which regulates the amount of water passing to the wheel, which is of the low-breast type. This shut consists of oak boards fastened to two curved oak arms which are operated by a rack-and-pinion gear turned from inside the building at both ground and first floor levels.

The wheel shaft drives the pit gear inside the building; this is 10 ft in diameter with 112 teeth, and the pit gear drives a horizontal gear known as the wallower which is 8 ft 6 inches in width with 96 teeth, thus giving a slight reduction. The wallower is keyed onto a round shaft which also drives an underdrift pan of 14 ft diameter. This shaft above the pan has a crown wheel enmeshing with a bevel gear which operates the hoist gear and two pumps, one for water, the other for pumping the slop material from the settling ark.

The hoist gear in the North Mill consists of two wooden sprockets with a slack chain which is tensioned by pulling onto a chain which lifts the top

North Mill gearwheels: the *Wallower* running in the horizontal plane, the *pit wheel* being driven directly by the waterwheel.

9
6
7
5
8
3
1
4
2
G.H.L.

Cheddleton North Mill

1 **Low-breast water wheel** is 22 ft diameter by 5 ft 9½ in wide. The arms are of oak, the paddle boards of larch

2 **Pit wheel,** so called because the portion below the axle runs in a pit. It is 10 ft diameter and has 112 teeth.

3 **Wallower,** the bevel gear driven in the horizontal plane which transmits the power to the pan. It is 8 ft 6 in diameter and has 96 teeth.

4 **Main shaft,** resting on a **toe step,** is round and, like the gears, of cast iron. The Wallower may be moved up the spline on the shaft in order to disengage it from the pit wheel.

5 **Grinding pan;** being driven from below is known as **underdrift.** It is 14 ft diameter, the **staffing** is of wrought iron. Refer to page 32 for details.

6 **Sweep arms,** of cast iron

7 **Hanging arms** of oak.

8 **Slug iron.**

9 **The crown wheel** enmeshes with a bevel gear which drives the horizontal shaft for the hoist. It also drives the water and slip pumps by means of a reciprocating arm: the pump inside supplied water for the pan, whilst that operated by a lever through the wall pumped the slop flint from the settling ark to the slip drying pan or kiln.

10 **Starts,** of seasoned oak, fit into boxes in the cast iron rim of the waterwheel.

11 **Floats,** or boards, of fresh larch, are bolted to the starts. They fit closely to the walls so that the maximum power may be obtained from the water.

12 **Fixed oak board,** secured to the walls.

13 **Sill.**

14 **Sluice,** or **shut;** oak boards are fastened to the two curved oak arms, which are raised or lowered to control the flow of water through the gap formed between the fixed board and the sill.

15 **Regulating gears,** consisting of toothed **racks** on the arms, enmeshing with **pinion** gears on a shaft controlled by wheels inside the mill building both on ground and first floors.

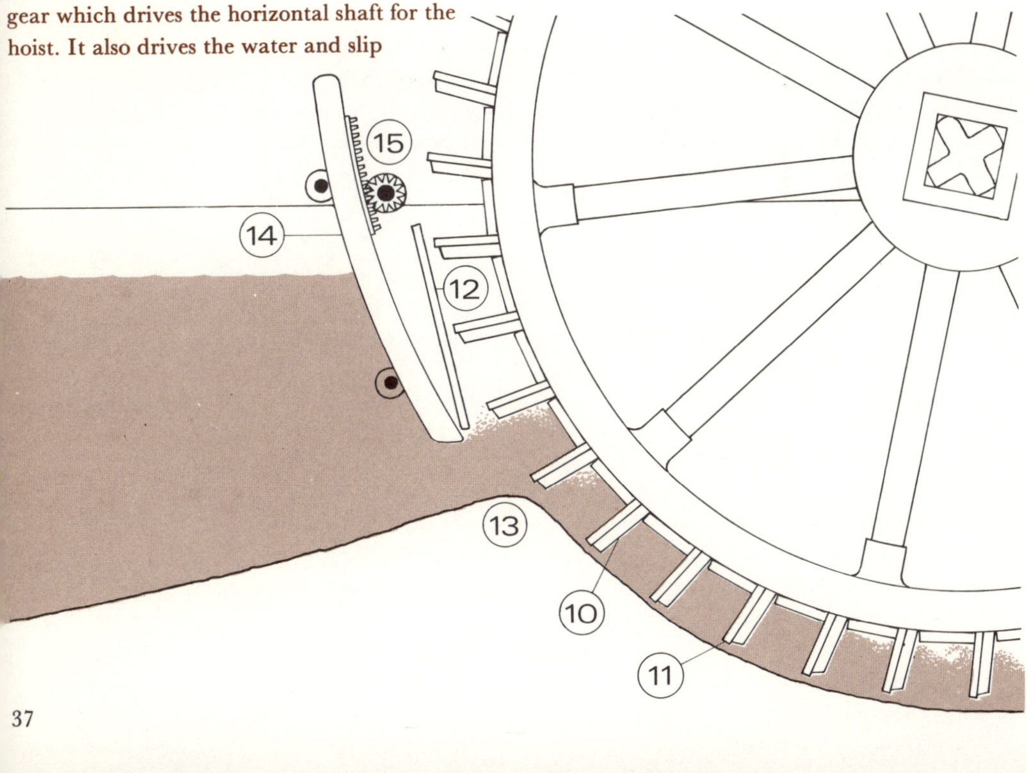

shaft, thus tensioning the chain on the sprockets to make the hoist lift. To lower the load there is a pawl in a ratchet which is lifted out of gear allowing the winch to lower its load.

The surface of the first floor is of brickwork, but from the hoist cast iron plates have been inserted into the floor to permit waggons to carry the heavy runner stones closer to the pan for lifting into position by the crane. This crane has been reconstructed from parts recovered from around the Mill during restoration. Also let into the floor are two channels from which the ground material is run either to the wash-tub and settling arks inside the Mill, or alternatively, as slop material to a trough which carried the material over the river to the South side where it would be processed in a similar way.

On the ground floor can be seen two large settling arks separated by a brick wall on which the wash-tub would have rested. The lean-to roof over these arks is supported at this point by a semi-circular section of an old corn-mill waterwheel rim into the centre of which would have been placed the upper end of the post of the wash-ark. In the right-hand settling ark there is a small area of the floor set slightly lower than the other levels which was for emptying the last residue of the material in that ark.

The South Mill

A study of the building shows that it has been altered and enlarged about three times. In the stone foundations by the wheel, there is evidence to show that a smaller wheel was originally used which may have been an undershot wheel, before the larger low-breast type was installed. The present wheel is 20 ft 5 inches diameter and 5 ft 5 inches in width. There are two sets of nine oak arms secured to the shaft by two cast iron hubs. The outer rim of the wheel is made up of cast iron segments, with boxes for the starts to fit in, as in the case of the North wheel. There are 36 L-shaped floats of larchwood, each 13 inches deep. The shaft, of different construction to that in the North Mill, is of octagonal cross-section and measures 16 inches across the flats; this drives the pit gear which is 6 ft 6 in diameter driving the horizontal gear, the wallower, which is also 6 ft in diameter. The pit gear has 89 teeth and the wallower has 88 teeth, the one tooth more on the pit gear being known as the hunting tooth: this is included to prevent the teeth meshing with the same one every revolution: this gives even wear on each tooth as the same two only come together every 88 revolutions.

The wallower is keyed onto a square shaft that drives an underdrift pan of 11 ft diameter. On the top of this shaft is a crown wheel forming the take-off drive to the two cast iron pumps; this gear may be disengaged by raising a carefully designed wooden arm which lifts it out of gear.

The hoist gear in this Mill consists entirely of wooden shafts and sprockets. The power take-off is a chain passing through two holes in the floor, and going down to a cast iron sprocket off the wheel shaft, between the wall and the pit gear. Because of the distance between the sprockets and the hoist, it needs two shafts of wood with 1:1 gear to overcome the difficulty of obtaining one continuous shaft. This hoist was originally installed when the Mill was still grinding corn.

The South Mill wheel

Additional machinery and exhibits

Since the Flint Mills were opened to the public, many items relating to the preparation of pottery raw materials have been assembled in order that the mill, besides being a preserved water mill typical of the period, may achieve its objective of becoming an industrial museum able to relate the history of pottery milling as completely as possible. Most of these items are to be seen in or around the South Mill, while the North Mill has been preserved as nearly as possible as it would have been in about 1815.

External exhibits

The exhibits include examples of vehicles for transporting the potters

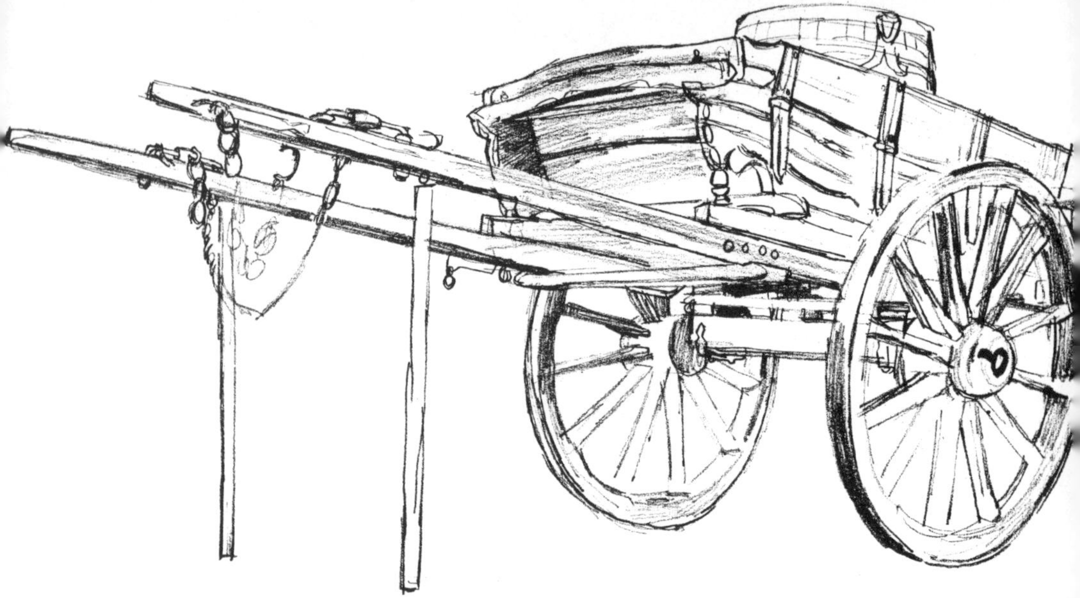

Left: Staffordshire one ton cart
Below left: View of mills from upstream April 1969.
Below: Tank used for carrying material in slop form.
Below right: Waggon with cast iron wheels.

materials to and from the mill.

Horse-drawn cart

When roads were good enough to allow wheeled carts and waggons to be used, materials were carried to and from the mill in them. Slop, or fluid, material would travel in casks or barrels, and later in tanks, any of which could be lifted off the cart. The one ton, two wheeled cart was the type most commonly used by potters millers. This cart at the mill is of the Staffordshire type.[108]

Plateway (tramway)

The earliest cast iron plates made to form a tramway were cast at Coalbrookdale in 1767. Those at Cheddleton Mill on which the old waggon stands are of the earliest type and may date from c 1777 when the Caldon Canal was opened: they were found in place. Other sections, excavated from the flint kilns, are of a later pattern: some stone blocks used to mount the plates (instead of stone or wood sleepers) come from the Coombe's Valley plateway running from the Caldon Canal to Mixon[167] where there was a copper mine. The waggon is dated 1803 (carved on the axle tree) and comes from the Spode Works in Stoke-on-Trent, which used similar plateways for horse drawn waggons to carry the clay to the potters workshops and materials from the canal side wharf to the mill and sliphouse.

Canalside crane

Used for unloading the heavy 'runner' stones from the canal narrow boats: brought from Goodwin's mill in Lichfield Street, Hanley where it was beside

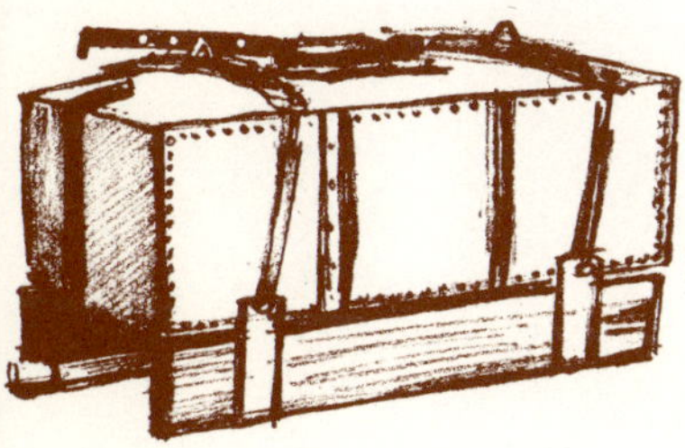

the Caldon Canal. It was re-erected on the site of an earlier crane.

Stone crusher

In many Mills the calcined flints would be crushed before grinding. Stamp Mills were used until superseded by jaw crushers or 'granulators'. Cornish stone always had to be crushed to a size suitable for grinding.

Centre feed grinding mill

Originally used for grinding pepper, it was brought from the East India Docks in London to Messrs Potclays Ltd of Etruria who used it for the fine grinding of china clay and other ceramic materials.

Edge runner mill

A very important class of mill used extensively wherever materials need to be ground. This small mill consists of two revolving stone edge 'runner' stone wheels rotating over a fixed stone 'bedstone' mounted in an iron frame. Another type consists of revolving runner wheels which remain in the same plane while the bedstone revolves beneath them. This small unit was used for grinding colour.

Exhibits in the South Mill, downstairs

Newcomen-type pumping engine model

Those manufacturers who expanded their business to any size preferred to control their own milling. Where there was no flow of water either they could build a windmill (as Thomas & John Wedgwood and John Baddeley did), or they installed a steam engine. Josiah Wedgwood bought a Watt beam

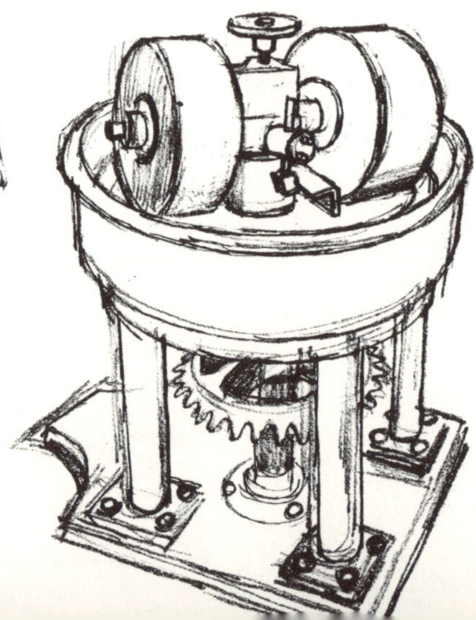

Above: Blake's patent stone breaker c1870 made by H. R. Marsden, Soho Foundry, Meadow Lane, Leeds.
Far left: Centre feed grinding mill.
Left: Edge runner mill.

Model of steam pumping engine.

engine in 1784 the first rotative engine in the Potteries.

But Josiah Spode, in his factory at Stoke was the first to use steam power: he used a Newcomen-style pumping engine to raise water to an overshot water-wheel as early as 1779. (See page 17).

This model is based on a design of James Watt (plans supplied by the Science Museum, London) which was typical of this method of providing rotative power before the rotative steam engine. Made for the Trust by the General Electric Company, Stafford and is of 'Old Bess' installed 1777 at Soho Works of Matthew Boulton & James Watt in Birmingham.

Reciprocating steam engine

A general purpose mill engine, built by Robey & Co. Ltd., of Lincoln and

Above: Reciprocating steam engine
Right: Single edge runner mill
Below: Frame sieve

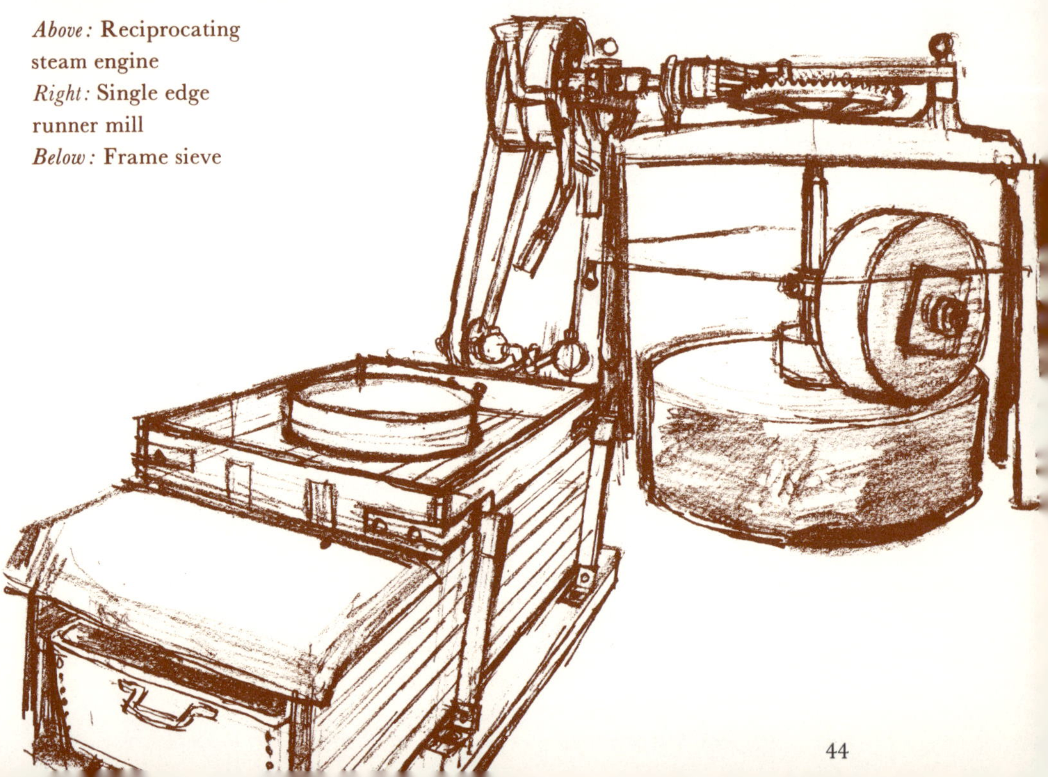

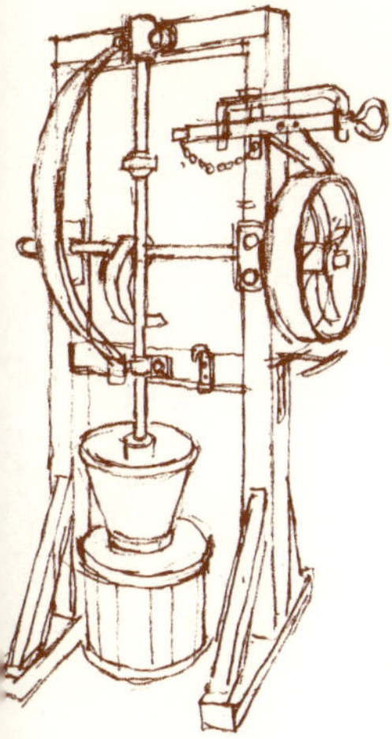

Single stamp mill

installed at Messrs Minton Ltd of Stoke on Trent, in 1928, to power their two-storey mill. The engine drove a steel shaft and catch-box from which vertical shafts then drove the grinding pans for bone, Cornish stone, glazes and colours. It replaced a beam engine and was one of the last engines to be installed for this purpose: generally steam engines were superseded by electric motors. This engine developed 100 hp and is fitted with the patent drop valve gear: the flywheel is 12 feet diameter and made in two sections which are bolted together.

This engine was brought to Cheddleton in May 1970.

Single edge runner mill

This mill was used for grinding colour. The 'bedstone' is of an unusual conglomerate stone, while the 'runner' is of granite. This type of mill was extensively used in preparing raw materials and also for blending them in the clay industries. Usually these mills were or iron and steel, with two revolving runners: the pan bottom sometimes revolved while the rollers revolved on fixed shafts.

Single Stamp mill

Adapted from the pestle and mortar. Stamp mills were in common use for crushing stone and flint (as well as various ore-bearing rock).[143] Before the introduction of wet grinding, stamp mills crushed flint and stone to the required fineness. Due to so many deaths and misery being caused by the silica dust, stamping was used latterly for crushing to small sized pieces before being ground on the wet pan. Multiple stamp mills may be seen in Cornwall. (A model of a set of Cornish tin stamps is on display).

Above: Steelyard and weighing-out box.
Right: Platform scales.

The shaft was raised by the eccentric cam lifting the shaft by operating on the fixed disc. (On permanent loan from Messrs. Wenger Ltd.)

Disintegrator, or granulator

A machine for pulverising substances. The material was subjected to the action of a vaned wheel revolving at high speed which flung the material against the stationary outer casing.

Frame sieve

A mechanical application of the simple sieve. This method of sieving was used before the introduction of vibrating sieves (lawns). It continued in use for certain substances, like chrome-tin colour, because it minimised the 'balling' of material which occurs on vibrating lawns. The scraper enabled the miller to break up any balls which did form.

Platform scales

Used for weighing material before mixing. (On permanent loan from Messrs. Wenger Ltd.)

Weighing-out box

Made of wood, these boxes were adjusted to a uniform weight by adding small pieces of lead. They were used for weighing materials for glaze and frit before melting in the frit kiln, or furnace. Wood was preferred because metal might have contaminated the substance. These boxes were weighed on a 'steelyard'.

In the North Mill there is a wooden barrow and steelyard used for the weighing of materials for bone china body mixing.

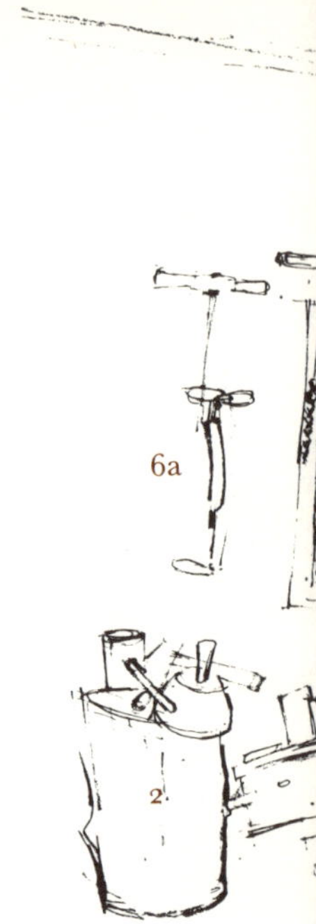

10

Exhibits in the South Mill, upstairs

Colour grinding pans or 'mullers'

The raw materials for ceramic colours are mixed as powders and then calcined in specially strong saggars to earthenware biscuit (1200°C) or glost (1060°C) temperature. The subsequent grinding is in small granite pans with a granite muller. These are of an early date, c.1800. (Given by Messrs. Doulton Fine China from the Minton Works, Stoke).

Glaze pan

Similar to normal pan but the staffing is made of wood to avoid the risk of metal contaminating the glaze. Also usually glaze pans are smaller than stone grinding pans.

Cask maker's tools

Casks were extensively used in the pottery industry both for the packing of finished ware and for transporting and storing raw materials of all kinds.

The local name for a cask-maker is a 'tub thumper' The 'chaffing' axe is designed with its handle offset to avoid the craftsman hitting his knuckles.

Also displayed are spoke shaves, mallet, and tub-thumper's block. (Given by George Edwards Esq.)

Wood turning lathe

Driven by foot treadle. This one, dating from about 1850, was used mainly for turning ark-plugs, the plugs used in the plug-plank of the settling ark. Also displayed are hand tools of the same period. (Given by George Edwards Esq., lately proprietor of Cheddleton Flint Mill).

1 Treadle lathe, with ark plug (1a)
2 Tub-thumpers block and tools
3 Wooden belt-pulleys
4 Barrel for dry colour
5 Wooden patterns for casting (a) South Mill water wheel rim section (b) spur wheel (unidentified)
6 Wheelwright's stool and some cart-making tools (6a)
7 Farrier's tools
8 Engineers tools, c1850
9 Stencils for marking casks, and box for individual stencils (9a
10 Plunger for clack pump

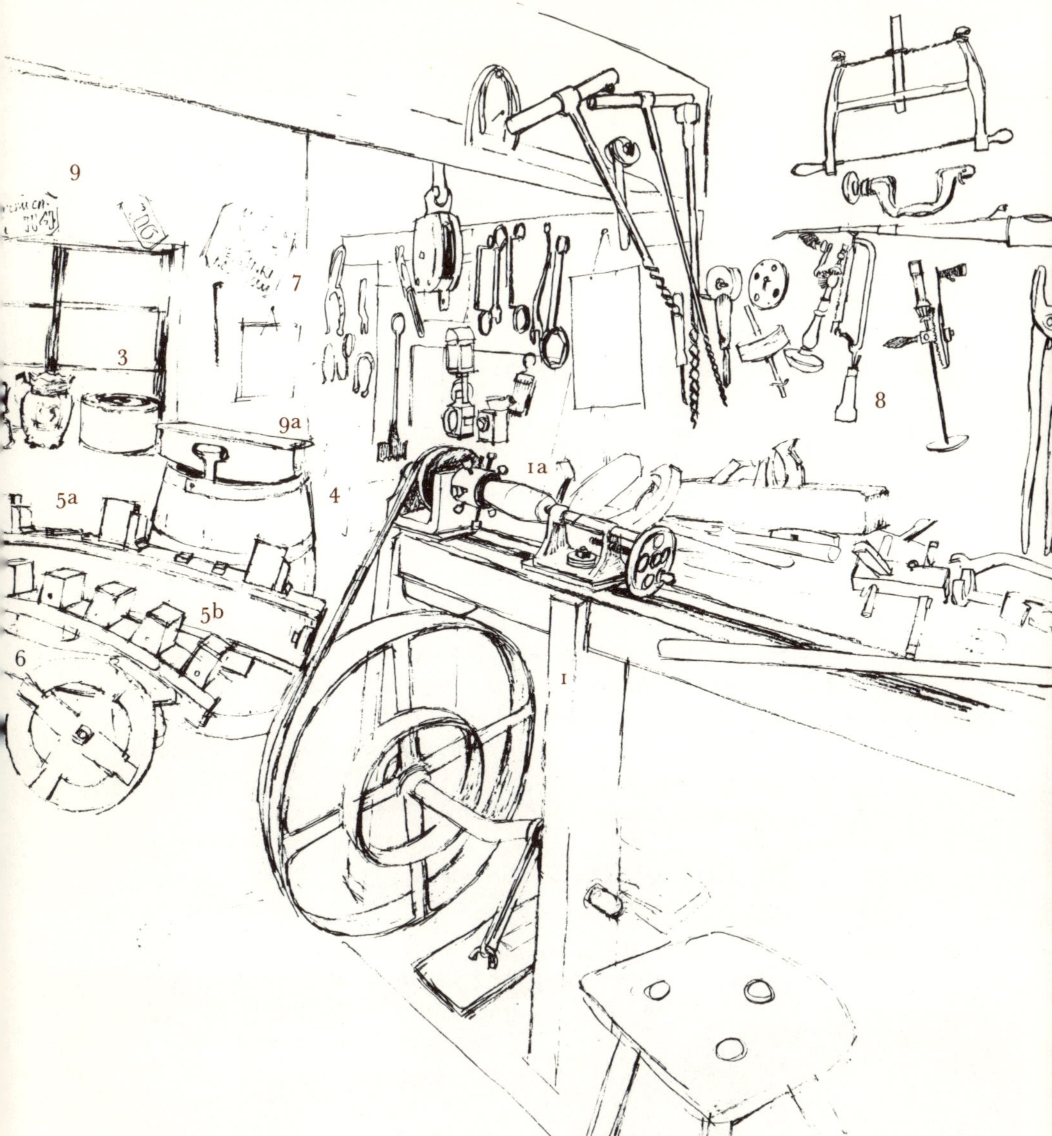
9
7
3
8
9a
1a
5a
4
5b
6
1

Appendix 1

Chronological list of some important developments in the pottery industry related to milling

Date	*Development*	*Source (page numbers)*
	Red Marl used since pre-Roman occupation	Rhead 10
1316	**Ironstone** mined in Tunstall	Ward 78
c1467	**Coal** (Grete Rowe): outcrops mined and used for firing kilns	Nicholls 7
c1615	**Coal** (Great Row) being mined	Ward 72
c1670	**Lead ore** applied to upper surface for glazing and the use of saggars introduced	Shaw 109
1671	**Salt:** Dwight introduces salt-glazing of stoneware at Fulham	Patent 164
c1680	**Salt** thrown on fires for glazing (introduction to N. Staffordshire)	Shaw 108
1685	**Fine Sand** (Mole Cop) mixed with clay (crouch ware)	Shaw 109–110
1694	**Salt:** duty imposed	Mountford 13
c1699	**Red ware:** Elers brothers fine quality potting (clay refined by levigation)	Solon 126
c1700	**Alabaster** Used for moulds	Mountford 30
c1720	**Chester clays** (ball clay brought from Devon) transported via Chester by packhorse as carts could not reach Burslem	Wedgwood 56
c1720	**Devon ball clays** quarried: introduction by John Astbury	Ward 48
c1720	**Calcined Flint** crushed by hand in iron mortar or by stamp mill: burnt with coal. Once believed to have been introduced to Staffordshire by Thomas Astbury; now known to have been in limited use before. Used in place of sand.	Ward 51 Nicholls 19 Weatherill 23
1726	**Wet grinding** Benson's Patent for grinding in water using iron balls in edge runner mill	Patent No. 487
c1730	**Slip kiln** introduced by Ralph Shaw to supersede sunpan for evaporating water	Wedgwood 58

Upper floor of a large grinding mill in the Staffordshire Potteries c1900

Date	*Development*	*Source (page numbers)*
1732	**Calcined flint:** Benson's second patent using stone balls in edge runner mill or pan mill	Pat. No. 536
1733	**Transport by river:** canalisation of River Weaver	Weatherill 90
1740	**Salt glazing:** dipping technique being used by W. Littler and Aaron Wedgwood	Shaw (Chemistry) 168
c1745	**Plaster of Paris** quarried as Gypsum and burnt: first recorded use for making moulds for shaping of pottery articles. Introduced by Ralph Daniel to replace alabaster (unburnt)	Plot 173 Ward 49
c1747	**China clay** and **Cornish stone:** deposits and properties of Cornish materials suitable for hard paste porcelain, discovered and recognised by William Cookworthy. The clay was dug (later washed) out, the stone quarried	Barton 18
c1749	**Calcined Bone** burnt with wood: first recorded use in a soft paste porcelain body at the Bow Factory, London	Burton 228
1750	**Lead oxide** in fluid suspension first used as fluid for transparent glaze by Enoch Booth	Wedgwood 67
1750	**Creamware:** casting of clay holloware with slip in moulds	Pococke 6
1759	**Roads turnpiked:** Act passed for road from Derby to Newcastle-under-Lyme	32 Geo. II Cap 60
1760	**Dorset blue ball clays** quarried and mined: Pike Brothers establish clay mines in Isle of Purbeck	
1762	**'Queen's Ware'** introduced by J. Wedgwood	Ward 430
1766	**Water transport:** Act of Parliament authorising Wilden Ferry (Trent) to Runcorn Gap (Mersey) Canal, 3 May	6 Geo III Cap 96
1766	**Grand Trunk Canal** (Trent & Mersey) begun by J. Brindley, 26 July	Boucher 94 Ward 170
1767	**Rail Transport:** first cast-iron waggon rails made at Coalbrookdale	Coalbrookdale Co. Ltd.
1768	**China clay** and **Cornish stone:** formula for porcelain patented by Cookworthy	Pat. No. 898
1772	**Water transport:** Trent & Mersey Canal length from Stone to Wilden Ferry opened	Mountford 55
1775	**China clay** and **Cornish stone:** Parliamentary struggle by J. Wedgwood against Champion's application for an extension to Cookworthy's patent. Succeeded in obtaining materials for general use outside limits of the patented formula. Act of Parliament	Thomas 15 (1936) 15 Geo. III Cap 52
	Salt: use for glazing discontinued almost entirely	Wedgwood 81
1776	**Water transport:** Act of Parliament authorising Caldon Canal	16 Geo III Cap 32
1777	**Water transport:** opening of North Staffordshire length of the Trent & Mersey Canal and Caldon branch, May	Ward 170
c1779	**First steam-powered engine** in the pottery industry (Newcomen-type to power waterwheel for flint and stone grinding) reported in the factory of J. Spode	Aikin 522 Hassall 62
1782	**Watt steam engine** (beam engine to drive grinding mills) installed at Etruria by J. Wedgwood	Thomas 13–25

Date	*Development*	*Source (page numbers)*
c1796	**Calcined bone** ground in water on pan: successful formula for Bone China discovered by Josiah Spode	Shaw 218 Burton 228
1796	**Calcined Flint** of medium grain size used for 'bedding' bone china in the biscuit firing	
1796	**Borax** introduced from Tibet for use in glazes	Wedgwood 157
1795–6	**Water transport:** Newcastle branch of canal opened, passing between factories of Spode & Wolfe and past Mintons	
c1796	**Ball clay mining:** Wedgwood, on behalf of the Potter's Clay Company, attained lease of ball clay setts in Isle of Purbeck	Barton 36
1799	**China clay mining:** Minton obtained interest in Treldar Common, and later formed the Hendra Company to work the china clay and stone.	Barton 36
	Spode and Wolfe lease Carloggas (1799–1820)	Barton 37–8
1800	**Tabernor's Rock** (ground): use of white burning argillaceous rock patented by J. & W. Turner	Pat. No. 2367
1813	**Ironstone clay,** in very small amounts, introduced into earthenware body; patented by C. J. Mason	Pat. No. 3724
1821	**Felspar** introduced by Spode into Porcelain body	Shaw 218
1815	**Rail transport:** tramway from Longton and Fenton to Stoke Canal Wharf	Wedgwood 163
1823	**Leadless glaze** introduced for earthenware by Job Meigh	Ward 383
1826	**Coal Gas** introduced into Burslem	Wedgwood 165
1837	**Rail transport:** Grand Junction Railway from Birmingham to Manchester opened	Wedgwood 169
1890	**Water:** piped supply starts	Wedgwood 165
1856	**Filter Press** for extracting water from clay slip under pressure produced by Needham & Kite of London. First seen at Messrs. Copelands factory	Lambert 63
1894	**Fritted Lead** introduced into glaze to reduce solubility and so minimise lead poisoning	Owen 297
1896	**Cylinder grinding:** Alsing licences William Boulton to manufacture cylinder mill	Fahey[164]
1900	**Lead Oxide (raw):** use prohibited by Home office	Owen 301

Drying kiln heated by underfloor flues (by courtesy of Jesse Shirley & Sons Ltd., Etruria).

Appendix 11

Some Patents relative to pottery milling and body development

Selected from a list in the first edition of Jewitt *Ceramic Art of Great Britain* (1876)

Date	*No.*	*To whom granted*	*Object*
1671	164	Dwight, John	Transparent earthenware and Stoneware (vulgarly
1684	234	Dwight, John	called Cologne-ware)
1722	448	Holt and London	Composition for Whiteware
1722	452	Billin, Thomas	Refined Earthenware from native materials
1726	487	Benson, Thomas	Grinding of Flint
1729	510	Bell, Samuel	Red marble ware
1732	536	Benson, Thomas	Improvements in grinding flint
1733	541	Shawe, Ralph	Chocolate and Whiteware, glazed with salt
1743 (Feb 10) *Provisional only*			
		Briand, Thomas[92]	Transparent Earthenware
1743 (Feb 28) *Provisional only*			
		Steers, William	Transparent Earthenware
1744	610	Heylyn and Frye	Manufacture of China
1749	649	Frye, Thomas	Manufacture of China using calcareous matter
1764	821	Williamson and Spackman	China
1766	849·	Lauraguais, Count de	China
1768	898	Cookworthy, William	China using Cornish Clay and Stone
1800	2367	Turner, W. and J.	Manufacturing of a fine Stoneware
1807		Spershott, James	Improvements in body
1810		Docksey, William	Preparation of glazing materials
1813	3724	Mason, C. J.	Improvements in composition of Earthenware
1845	10968	Skinner and Whalley	Manufacture of pottery bodies, compositions etc.
1845	11005	Findler, Thomas	Improvements to edge runner mills for grinding
1846	11313	Fourdrinier, G. H.	Frame sieve for separating particles
1849	12789	Browne and Veale	Grinding Flint
1854	1933	Mayer and Bush	Grinding Flint
1855	97	Hollins, M. D.	More efficient use of fuel and design of slip kilns
1855	208	Mayer and Bush	Grinding Flint
1856		Needham and Kite	Filter Press for expressing liquids

Date	*No.*	*To whom granted*	*Object*
1857	997	Harland, John	Purifying clay
1864	1024	Worssam, G. J.	Separation of solids in slip
1864	1576	Cockran, R.	Treating clay
1865	805	Wright, J.	Dewatering slip by centrifugal action
1867	1605	Orr, William	Drying potters clay
1867	1707	Orr, William	Separating liquids from substances
1867	1939	Borlase, Thomas	Machine for grinding ores
1867	2065	Fletcher, Henry	Mixing and drying clay in rotary kilns
1867	2975	Abel, C. D., for Alsing	Pulverisation of materials in Ball Mill
1867	3319	Boulton, William	Transmission of power by endless rope
1867	3577	Kerr, William H.	Calcination of materials for pottery
1873	2026	Bodmer and Bodmer	Dewatering clay by rollers and rotary kiln
1874	1119	Boulton, William	Improved blunger with angled blades and vibrating sieves
1874	2537	Alsing	Improvements to Patent No. 2975 of 1867
1874	3076	Scott, W. S.	Separating water from slip: avoidance of pumpwear
1875	1745	Hill, J.	Separating water from slip: primitive pug

First floor of North Mill, Cheddleton

References and Bibliography

The list of books and publications is comprehensive so that students and researchers may benefit as much as possible. Some of the books listed are not referenced to the text, but are included here because valuable information may be obtained from them; also some cross-references are included to avoid too many numerals in the main text.

1 Aikin, J. (1795) *A description of the Country from thirty to forty miles round Manchester*. London: p.116.

2 Ibid, p.522.

3 Ibid, pp.526–7.

4 Anon. (1960) *The Clay Mines of Dorset – worked by Pike Bros. Fayle & Co Ltd of Wareham 1760–1960*. London: p.10.

5 Anon. (1966) Grinding for the pottery trade. *Edgar Allen News*. Vol. 45, No. 526, pp.85–87.

6 Ibid, No. 527, pp.110–111.

7 Anon. (1967) Victoria History of the Counties of England. *A History of the County of Stafford*. Oxford. Vol. II, pp.5–6.

8 Barnard, H. (1920) *Artes Etruriae Renascuntur*. Stoke on Trent.

9 Barton, R. M. (1966) *A History of the Cornish China Clay Industry*. Truro: pp.18–19. See Ref. 34.

10 Ibid, pp. 21–23.

11 Beaver, S. H. (1964) The Potteries: A study in the evolution of a cultural landscape. *Trans. Inst. Brit. Geog.* Publication No. 34, pp.6–7.

12 Bimson, M. (1961) John Dwight. *Trans. Eng. Cer. Circle.* Vol. 5, Part 2, pp.95–109.

13 Bladen, V. W. (1926) The Potteries in the Industrial Revolution. *The Economic Journal:* pp.121–2.

14 Ibid, p.124.

15 Boucher, C. T. G. (1968) *James Brindley, Engineer, 1716–1772*. Norwich: p.18.

16 Ibid, p.89.
17 Boucher, C. T. G. (1966) The Schemer. *Advance*, University of Manchester: Vol. I. p.25.
18 Bourry, E. (1911) *A Treatise on Ceramic Industries* (trans. & rev. A. B. Searle). London: p.60.
19 Ibid, pp.92–93.
20 Ibid, pp. 476–490.
21 Boulton & Watt Mss. Birmingham Public Libraries (Reference Library).
22 Brears, P. C. D. (1971) *The English Country Pottery: its history and techniques*. Newton Abbott.
23 Bulley, J. A. (1955) The Beginnings of the Devonshire Ball Clay Trade. *Trans. Dev. Assn.* Torquay. Vol. LXXXVII, p.192.
24 Ibid, p.195.
25 Ibid, pp. 203–4.
26 Burton, W. (1906) *Porcelain, A Sketch of its nature and manufacture.* London: p. 228.
27 Carruthers, R. G. and Strahan, A. (1923) *Lead and Zinc Ores of Durham, Yorkshire and Derbyshire.* (Special Report on the Mineral Resources of Great Britain). Memoirs of the Geological Survey, London. Vol. XXVI, pp.41–8.
28 Celoria, F. (1971) Reprint of Dobson, E. (1850) *A Rudimentary Treatise on the manufacture of bricks and tiles. Journal of Ceramic History*, No. 5. Stafford. Part I, p.6.
29 Champion, R. (1775) Extension of Cookworthy Patent assigned to Champion. Act of Parliament 15 Geo. III Cap. 52. See also Jewitt, L. articles in The Art Journal Sept., Nov. & Dec. 1863.
30 Chapman–Huston, D. and Cripps, E. C. (1954) *Through a City Archway. The story of Allen and Hanburys 1715–1954.* London: p.30
31 Charles, B. (1972) Mss of *Glossary of Pottery Terms*. See Ref. 4.
32 Clark, R. H. (1963). The Steam Engine in Industry and Road Transport. *Engineering Heritage*, London: Vol. I. p.63
33 Cock, D. (1880) *A Treatise Technical and Practical on the nature, production and uses of China Clay*. London: p. 123.
34 Cookworthy, W. (1768) Patent No. 898
35 Cornforth, J. (1966) Profits, Patronage and Politics: The Lunar Society. *Country Life*, Oct 20, pp.968–970
36 Dale, O. A. (1968) Personal Communication
37 Dodd, G. (1844–5) *Account of the manufactures of Great Britain.* London: Vol. II p.210
38 Dodd, A. E. and E. M. (1963) The Froghall-Uttoxeter Canal. *N. Staffs Jl. Field Studies*. Vol. 3, p.59
39 Evans, W. (1846) *Art and History of the Potting Business*. Stoke on Trent
40 Ford, T. D. (1972) Personal Communications. Robert Thornhill owned a quarry at Great Longstone in Derbyshire, and his account book for the years between 1796 and 1881 show that between 1796 and 1811 some 3,500 tons of chert were supplied to various potters in

Canalside crane. Davenport Grindrod made at Caledonian Foundry, Liverpool quantities ranging from 4 to 50 tons per annum. Most of the accounts specify only chert stone, but two note 'pavers', and one lists 'runners, pavers and setters' at £1.3.6, £1.5.0 and £1.5.0 respectively. Two accounts give details of deliveries to Mr. Thomas Cheadle and Mr. William Adams at Cheadleton. See Ref. 169

41 German, W. L. & Singer, F. (1960) *Ceramic Glazes*. London: p. 13

42 Ibid, p.20

43 Green, A. T. (1930) The Contributions of Josiah Wedgwood to the Technical Side of the Pottery Industry. *Trans. Cer. Soc.*, Vol. XXIX, p.24

44 Greenslade, M. W. & Stuart, D. G. (1965) *A History of Staffordshire*. Beaconsfield.

45 Haggar, R. G. (1952) *The Masons of Lane Delph and the origin of Masons Patent Ironstone China*. London: p.14

46 Haggar, R. G. (1966) *Pottery through the Ages*. London.

47 Haggar, R. G. (1972) Personal Communication. 'In 1799 import duties on porcelain from the Far East had reached the astonishing figure of £109.8.6d for every £100 value of ware imported without any drawback. *Statutes at Large*, Vol. 18 (1799) Ch. 59, Schedule A.' Refers to Falkner, F. (1912) *The Wood Family of Burslem*: p.67. See Ref. 45

48 Hargreaves, T. (1832) *Map of Stoke on Trent*

49 Hassall, J. (1819) *A Tour of the Grand Junction Canal in 1819*. (Republished 1968). London: p.62

50 Helsby, L. E., Rushton, A. J. & Legge, D. R. (1961) *Watermills of the Moddershall Valley – An Historical Engineering and Economic Survey of the Water Mills of Staffordshire*. A project for the College Diploma Course at Stafford College of Technology.

51 Howe, J. A. (1914) *Handbook to the Collection of Kaolin, China-Clay and China-Stone in the Museum of Practical Geology*. London: p.137

52 Hudson, K. (1969) *The History of English China Clays*. Newton Abbott

53 Hunt, R. (1881) *Mineral Statistics of the United Kingdom*. Mem. Geol. Survey, p.141

54 Hutchins (1861) *History and Antiquities of the County of Dorset*. London: p.95

55 Inquisition 37 Hen III 1253

56 Jewitt, L. (1883) *The Ceramic Art of Great Britain*. 2nd Ed. London: p.519

57 Kirkham, N. (1971) Personal Communication. Magnus – cites Agricola, G. (1556) *De Re Metallica*. 'Magnes —magnetite—loadstone.' It seems that the 'magnus' mentioned by Dr Plot as being used in glazing was most likely to have been magnetite (magnetic oxide of iron Fe_3O_4) with perhaps some manganese associated with it. See Ref. 27.

58 Ibid. Galena – cites i) Farey, J. (1815) *A General view of the Agriculture and Minerals of Derbyshire*. Vol. I, p.378. 'Formerly it seems that considerable quantities of the Derbyshire Lead Ore, selected

from the largest and most pure, or Bing Ore, was sold in that state to the Staffordshire and other Potters, to grind, in preparing the glazes for their common wares, and hence was called Potter's ore;' . . . Galena – cites ii) Mander, J. (1824) *Derbyshire Miner's Glossary*, p.48. . . . 'Potter's Ore is the best; it is soft and tender, and easily broken . . .' Miss Kirkham believes that references regarding Potter's Ore would be to Bing Ore. This was good quality ore, mined in fair-sized lumps with little rock or other minerals adhering to it, so that it was easy to dress. Smitham (sic) was small powdered ore which resulted from sieving and washing: up to the sixteenth century it was thrown away, but better smelting processes enabled it to be used after then. The lead miner would not have dressed his ore to Smitham purposely: it paid him better to sell Bing ore to the potters who would have done their own smithaming. See Refs. 65, 66, 104, 105

59 Ibid. Pipe-clay – cites Pilkington, J. (1784) *A View of the present state of Derbyshire*. Vol. I, p.162. 'Porcelain clay of a most delicate white colour, and a very fine texture from a lead mine near Brassington.' See Ref. 168, p.163

60 Lambert, G. (1865) *Art Ceramique. Description de la fabrication actuelle des Faiences fine et autres potteries en Angleterre*. Brussels: pp.32–3

61 Ibid, p.40 *et seq.*

62 Ibid, p.68

63 Mankowitz, W. (1967) *Wedgwood.* London: (Re-print of 1953 Ed). London p.45. See Ref. 56, p.511

64 Mankowitz, W. & Haggar, R. G. (1957) *The Concise Encyclopaedia of English Pottery and Porcelain*. London:

65 Mawe, J. (1802) *The Mineralogy of Derbyshire*. London: p.42

66 Ibid p.209

67 Meredith, W. D. (1964) Watermills in North Staffordshire. *N.Staffs Jl. Field Studies*, Vol. 4, p.1

68 Meteyard, E. (1866) *Life of Josiah Wedgwood* London: (1970 Reprint), Vol. I, pp.139–140 Meteyard states that 'balls' or lumps of clay weighed 60–70 pounds each, and each horse carried six balls – three in a pannier on either side. See Ref. 110

69 Ibid, p.149

70 Ibid, p. 152. See Ref. 77

71 Ibid, p.156. See Ref. 152

72 Ibid, p.157. Cites Knapp, F. (1848) Chemical Technology, or *Chemistry applied to the Arts and Manufactures*, Vol. II, pp.475, 477–8

73 Ibid, p.159

74 Ibid, p.181. See Ref. 77, 137, 129

75 Ibid, p.215. See Ref. 83.

76 Ibid, pp. 266–7.

77 Ibid, pp.279–80. See Ref. 131

78 Ibid, p.436

79 Ibid, p.453

80 Ibid, Vol. II, p.242

81 Mountford, A. R. (1971) The Illustrated Guide to *Staffordshire Salt-Glazed Stoneware*. London: p.4
82 Ibid, p.5
83 Ibid, p.11
84 Ibid, p.21
85 Ibid, p.25
86 Ibid, p.26
87 Ibid, p.27
88 Ibid, p.36. See Ref. 12
89 Ibid, p.39
90 Ibid, p.55
91 Ibid, Personal Communication
92 Mountford, A. R. (1969) Thomas Briand – A Stranger. *Trans. Eng. Cer. Circle*. Vol. 7, Part 2, p.87
93 Nicholls, R. (1931) *Ten Generations of a potting family*, London p.19
94 Ibid, p.86
95 Orme, G. R. and Ford, T. D. (1970) Polyphase Mineralization in Chert from the Ashford Black Marble Mine, Derbyshire. *Proceedings of the Yorkshire Geological Soc.*, Vol. 38, Number 8
96 Owen, H. (1901) *The Staffordshire Potter* London, p.297
97 Ibid, p.301
98 *Penny Magazine* (1843) *A Day at the Staffordshire Potteries*. Vol. XII, No. 716, pp.201–208
99 Pitt, W. (1817) *A Topographical History of Staffordshire: A Brief Account of the Staffordshire Potteries*. Newcastle under Lyme, p.416. See Ref. 165
100 Ibid, p.417
101 Ibid, p.435
102 Plant, H. J. (1939) Alumina and its uses in the china trade. *Trans. British Ceramic Soc.*, Vol. 38, pp. 476–490
103 Plot, R. (1686) *The Natural History of Staffordshire*. Oxford: p.121
104 Ibid, p.123. See Ref. 57
105 Ibid, p.166
106 Ibid, p.169–70
107 Ibid, p.173
108 Ibid, p.354
109 Pococke, R. (1750) *The Travels through England of Dr. R. Pococke*. pp.6–8
110 Polewhele, R. (1806) *History of Devonshire. Fossils, Native and Extraneous*. pp.37–8. See Ref. 127
111 *Pottery (Health) Special Regulations* (1947) (a) SR and O 1947, No. 2161. Sec.5(a)
112 *Ramsden Bulletin* (1971) Water Mills for Grinding Minerals: Stoke on Trent. No. 382, May 1971
113 Ibid, No. 385, August 1971
114 Ibid, No. 386, October 1971
115 Rhead, G. W. and F. A. (1906) *Staffordshire Pots and Potters*. London. p.146. See also Rix W. P. (1901) Notes on Vaporous Method of Glazing Pottery. *Trans Eng. Cer. Soc.* Vol. 1. pp. 26–35.

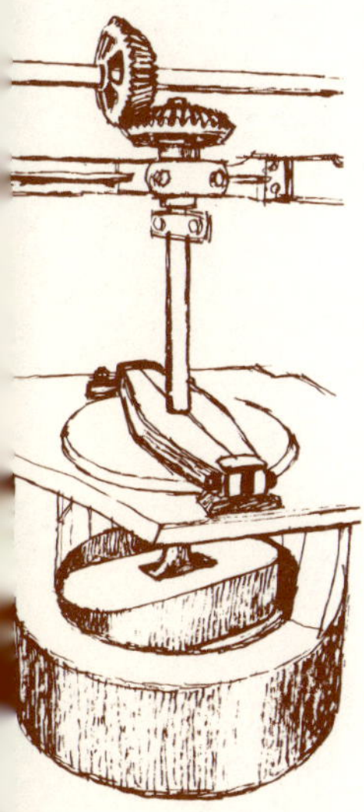

Colour grinding pan using granite bedstone & muller

116 Ibid, p.149
117 Ibid, p.150
118 Ibid, p.182. See Ref. 84
119 Rosenthal, E. (1949) *Pottery and Ceramics* from common brick to Fine China. London: p.73–5
120 Sandeman, E. A. (1901) *Notes on the Manufacture of Earthenware.* London:
121 Scott, A. (1929) *Ball Clays* Memoirs of the Geological Survey, London: Vol. XXXI.
122 Shaw, S. (1829) *History of the Staffordshire Potteries.* Hanley (1901–Reprint): p.63. See Refs. 123, 128, 89
123 Ibid, pp.97–8
124 Ibid, p.110. See Ref. 99
125 Ibid, p.125. See Ref. 13
126 Ibid, p.147. See Refs. 73, 14
127 Ibid, p.149. See Ref. 68
128 Ibid, p.157
129 Ibid, p.160
130 Ibid, p.173. See Refs. 2, 145
131 Ibid, p.178. See Ref. 77
132 Shaw, S. (1837) *Chemistry of Pottery.* London: p.245
133 Ibid, pp.247–8
134 Sherlock, R. L. (1921) *Rock Salt and Brine.* Special Report on the Mineral Resources of Great Britain, London: Vol. XVIII. p.2 See Geological Survey Cheshire. p.262
135 Sherlock, R. J. (1970) Personal Communication (2 March 1970). Draft entry for Brewery and Silk Factory, Cheddleton, *The Industrial Archaeology of Staffordshire*
136 Smiles, S. (1862) Lives of the Engineers with an account of their Principal Works . . . London: Vol. I. Part v. Life of James Brindley. p.328
137 Ibid, p.439. See Ref. 79
138 Ibid, p.470. See Refs. 80, 101
139 Solon, M. L. (1885) *The Art of the Old English Potter.* 2nd Edition. London.
140 Thomas, A. L. (1934) *Geographical aspects of the Development of Transport and Communications affecting the Pottery Industry of North Staffordshire during the Eighteenth Century.* Manchester (University Thesis), [Collections for a history of Staffordshire 1934] p.58
141 Ibid, pp.80–89
142 Thomas, J. (1936) Josiah Wedgwood as a pioneer of steam power in the pottery industry. *Trans. Newcomen Soc.*, Vol. 17, pp.13–25
143 Thomas, J. (1971) *The Rise of the Staffordshire Potteries* (based on thesis of 1934). Bath: p.45. See Refs. 145, 8
144 Ibid, p.47
145 Thomas, J. (1933) Josiah Spode – his Times and Triumphs. *Trans. Ceramic Society*, Vol. XXXII, pp. 489–512
146 Wailes, R. (1966) Water Driven Mills for Grinding Stone. *Trans. Newcomen Soc.*, Vol. XXXIX, p.105

Glaze pan
with wood staffing

147 Ibid, p.110
148 Ward, J. (1843) *The Borough of Stoke-upon-Trent* . . . comprising its history etc. London: p.52. See Ref. 132, p.417
149 Ibid, p.165
150 Ibid, pp.593–4
151 Weatherill, L. (1971) *The Pottery Trade and North Staffordshire 1660–1760* Manchester. pp.15–16
152 Ibid, p.17. See Ref. 71
153 Ibid, p.18
154 Ibid, p. 20. See Refs, 123, 22
155 Ibid, p.23
156 Ibid, p.27. See Ref. 159
157 Ibid, p.88 *et seq*. See Ref. 163
158 Ibid, p.90
159 Ibid, p.157
160 Wedgwood, J. C. (1913) *Staffordshire Pottery and its History*. London: p.67
161 Whiter, L. (1970) *Spode:* a history of the family, factory and wares from 1733. London: p.27
162 Yates, W. (1775) *Map of the County of Stafford from an actual Survey 1769–1775*
163 Young, A. (1770) *A Six Month's Tour through the North of England*. London. Vol. III, pp.306–312
164 Fahey, D. (1972) Personal Communication, and details from William Boulton's Catalogue (1900) and personal notebook of Francis G. Harris (1896–8). The first patent was registered in 1867 (No. 2975), with an improved version being patented in 1874 (No. 2537). Alsing entered into an agreement with William Boulton in 1896 to manufacture and market his Pulverising Cylinders in the United Kingdom, and many were sold in that year. But before the agreement was ratified, Alsing 'had put in two cylinders for Mr. Wenger at a very reduced price . . .'. It seems that Alsing had offered Wenger an agency to sell cylinders made in Europe for Alsing before negotiating the agreement with Boulton. William Boulton's catalogue of 1900 illustrates nine different models of cylinders.
165 Evans, W. B., Wilson, A. A., Taylor, B. J., and Price, D. (1968) *Geology of the Country around Macclesfield, Congleton, Crewe and Middlewich*. Memoirs of the Geological Survey. London, pp.68–69. See Refs. 99,106
166 Ibid, p.262. See Ref. 134
167 Green, A. and Strahan (1887) *The Geology of the Carboniferous Limestone, Yoredale rocks and Millstone Grit of North Derbyshire*. Memoirs of the Geological Survey. London, p.158
168 Ibid, p.159. See Ref. 104
169 Ibid, p.166. See Ref. 40

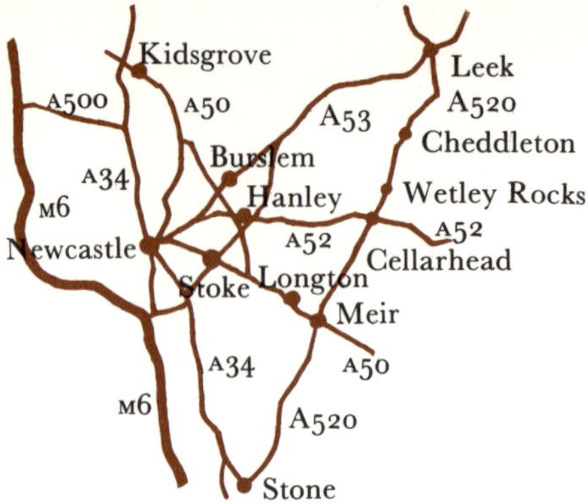

The Cheddleton Flint Mill Preservation Trust was formed in 1967 to preserve the mill and provide educational information concerning the historical development of pottery raw materials.

Now, in 1972, the Trust is widening its objectives to embrace more of Britain's Industrial Heritage.

The mill is open on Saturday and Sunday afternoons throughout the year.

Admission is free, but gifts of money are very necessary if the Trusts objectives are to be achieved.

The Trust is registered as a Charity.